International Perspectives in Geography
AJG Library 3

More information about this series at http://www.springer.com/series/10223

Aims and Scope:

The AJG Library is published by Springer under the auspices of the Association of Japanese Geographers. This is a scholarly series of international standing. Given the multidisciplinary nature of geography, the objective of the series is to provide an invaluable source of information not only for geographers, but also for students, researchers, teachers, administrators, and professionals outside the discipline. Strong emphasis is placed on the theoretical and empirical understanding of the changing relationships between nature and human activities. The overall aim of the series is to provide readers throughout the world with stimulating and up-to-date scientific outcomes mainly by Japanese and other Asian geographers. Thus, an "Asian" flavor different from the Western way of thinking may be reflected in this series. The AJG Library will be available both in print and online via SpringerLink.

About the AJG

The Association of Japanese Geographers (AJG), founded in 1925, is one of the largest and leading organizations on geographical research in Asia and the Pacific Rim today, with around 3000 members. AJG is devoted to promoting research on various aspects of human and physical geography and contributing to academic development through exchanges of information and knowledge with relevant internal and external academic communities. Members are tackling contemporary issues such as global warming, air/water pollution, natural disasters, rapid urbanization, irregular land-use changes, and regional disparities through comprehensive investigation into the earth and its people. In addition, to make the next generation aware of these academic achievements, the members are engaged in teaching and outreach activities of spreading geographical awareness. With the recent developments and much improved international linkages, AJG launches the publication of the AJG Library series in 2012.

Editorial Board:

Yoshiyasu Ida • Minori Yuda • Takashi Shimura
Shunsuke Ike • Koji Ohnishi • Hideki Oshima
Editors

Geography Education in Japan

Springer

Editors
Yoshiyasu Ida
University of Tsukuba
Ibaraki, Japan

Takashi Shimura
Joetsu University of Education
Niigata, Japan

Koji Ohnishi
Toyama University
Toyama, Japan

Minori Yuda
The University of Tokyo
Tokyo, Japan

Shunsuke Ike
Waseda University
Tokyo, Japan

Hideki Oshima
Keio University
Yokohama, Japan

ISSN 2197-7798 ISSN 2197-7801 (electronic)
ISBN 978-4-431-54952-9 ISBN 978-4-431-54953-6 (eBook)
DOI 10.1007/978-4-431-54953-6
Springer Tokyo Heidelberg New York Dordrecht London

Library of Congress Control Number: 2014955571

Printed on acid-free paper

Springer is part of Springer Science+Business Media (www.springer.com)

Preface

Since education in Japan is conducted in Japanese, this is a barrier for non-Japanese speakers who are interested to learn about instructional practices and problems that exist in the country. Even though there is interest in Japanese geography education, there are few opportunities to learn about it owing to this language barriers.

Geography education researchers in Japan have not actively published in English language journals. Therefore, researchers and teachers in other countries only have limited information about the status of Japanese geography education. Recently, many scholars have organized international projects, which have contributed to developing collaboration across nations. From this perspective, a book published in English on geography education in Japan is a timely contribution to the field.

This book provides a comprehensive description of Japanese geography education from diverse viewpoints. The geography education committee of the Association of Japanese Geographers (AJG) led the editing work that resulted in a volume that is part of the International Perspectives in Geography: AJG Library series.

This book includes the following broad topics: introduction to geography education in Japan, regional geography and field surveys as essential elements of geography education, geography education and cooperation with the local community, support for teachers from professional geographers and the AJG, international comparative studies on geography education, and the future of geography education in Japan. The editing team greatly appreciates the hard work of all contributors to this book. We also thank Dr. Niem Tu Huynh for carefully reading through each chapter to make comments and suggest changes.

We hope this book will contribute to the improvement of geography education worldwide.

Ibaraki, Japan — Yoshiyasu Ida
Tokyo, Japan — Minori Yuda
Niigata, Japan — Takashi Shimura
Tokyo, Japan — Shunsuke Ike
Toyama, Japan — Koji Ohnishi
Yokohama, Japan — Hideki Oshima

Contents

Contributors

Hiroaki Akimoto Dokkyo University, Saitama, Japan

Toshio Asakawa Saitama Prefectural Urawa Daiichi Girls' Senior High School, Saitama, Japan

Kaoru Hirasawa Shiraoka City Municipal Shinozu Junior High School, Saitama, Japan

Yoshiyasu Ida University of Tsukuba, Ibaraki, Japan

Shunsuke Ike Waseda University, Tokyo, Japan

Koji Ishizuka Tokiwa University, Ibaraki, Japan

Tomoaki Ito Shizuoka Prefectural Susono Senior High School, Shizuoka, Japan

Hiromi Iwamoto Nara University of Education, Nara, Japan

Takahisa Izumi Senshu University Matsudo High School, Chiba, Japan

Hyunjin Kim Hokkaido University of Education, Hokkaido, Japan

Tomoko Murayama Ibaraki University, Ibaraki, Japan

Shuichi Nakayama Hiroshima University, Hiroshima, Japan

Koji Ohnishi University of Toyama, Toyama, Japan

Hideki Oshima Keio University, Yokohama, Japan

Takashi Shimura Joetsu University of Education, Niigata, Japan

Jun'ichiro Takata Gifu Shotoku Gakuen University, Gifu, Japan

Hirokazu Takeuchi Chiba University, Chiba, Japan

Yumiko Takizawa Teikyo University, Tokyo, Japan

Kiyoshi Teramoto Tamagawa University, Tokyo, Japan

Katsuki Toida Kinki University, Osaka, Japan

Fumio Wada Fukuyama University, Fukuyama, Japan

Kotaro Yamagata Joetsu University of Education, Niigata, Japan

Kazuyoshi Yoshida Soka University, Tokyo, Japan

Hiroya Yoshimizu Hyogo University of Teacher Education, Hyogo, Japan

Minori Yuda The University of Tokyo, Tokyo, Japan

About the Authors*

Hiroaki Akimoto was born in 1962 in Saitama Prefecture. Akimoto graduated from the University of Tsukuba. He taught at Saitama public high school and the Tokyo Gakugei University Senior High School. Dr. Akimoto is currently a professor in the Economics Department at Dokkyo University in Saitama.

Toshio Asakawa is a teacher at Urawa Daiichi Girls' Senior High School, with 31 years of experience teaching geography to senior high school students in Saitama Prefecture. He worked as a Senior Curriculum Specialist at the National Institute for Educational Policy Research from 2004 to 2006, and also as a teaching consultant at Saitama Prefectural Board of Education from 2007 to 2008. He has published several books and papers, including *Impressive Geography Teaching Materials 4* (2009, in Japanese).

Kaoru Hirasawa served as a teacher in the Saitama University Faculty of Education's Attached Junior High School. Having served in the education administration of Kazo City and Saitama Prefecture, Hirasawa is presently principal of Shiraoka City Municipal Shinozu Junior High School. His specialty is geography education and he has published articles including "The various regions of Japan," jointly authored with Aoyagi Shin'ichi and Fujisaki Akitaka in Nakamura Kazuo et al. (ed) (2009). *A Course in Geographic Education Vol. III: Geography Education and Maps and Descriptive Geography*. Tokyo: Kokon Shoin.

Yoshiyasu Ida is a professor of Human Sciences at the University of Tsukuba. He teaches geography education and geography with a focus on the geography curricula of various countries, particularly those of Oceania including New Zealand, Australia, Micronesia and Palau. He has written "Japan: GIS-enabled field research and a cellular phone GIS application in secondary schools" with Minori Yuda

* The information on authors are as of the end of March 2014.

(Milson, A.J., Demirci, A. and Kerski, J.J. (eds.) (2012). *International Perspectives on Teaching and Learning with GIS in Secondary Schools*. (pp. 141–149). *Region and Education* (2005, in Japanese), as well as other books and papers.

Shunsuke Ike (Ph.D.) was born in Tokyo in 1959. After serving as a professor in the Faculty of Education at Shizuoka University, he is now a professor at the Faculty of Education and Integrated Arts and Science at Waseda University. He specializes in rural geography, regional geography of Portugal, and geography education. His main published books are *Tourism Use of Common Space in Rural Areas* (2006, in Japanese) and *Fifty-five Chapters for Understanding Portugal* (2011, in Japanese).

Koji Ishizuka is a professor in the College of Community Development at Tokiwa University. He has worked in the Institute for Areal Studies Foundation for 27 years, and then became a faculty member of the University in 2006. His major research field is human geography and regional policy. He has written "The Basic Study of Regional Policy" (*Community Development Studies*, 10, 2010), "A Study on the University Student's Motivation for Learning" (*Community Development Studies*, 14, 2012), and other related papers.

Tomoaki Ito is a teacher at Yoshiwara senior high school in Shizuoka Prefecture. He has taught high school geography for 17 years and has developed GIS materials for geography education since 2001. He is well known by geography teachers for his web site entitled "ITOCHIRI" (http://www.itochiri.jp/). He has written "ITOCHIRI-style GIS for geography classes" (2010, in Japanese).

Hiromi Iwamoto is a professor in the Social Studies Department at Nara University of Education. He taught elementary school for 12 years, and then became a faculty member of the University in 1991. His major research fields are geography education within social studies and life environment studies. He has written "Social studies education research using Okinawa as a case study" (*Bulletin of the Center for Educational Research and Development*, 21, 2012, in Japanese), and related books and papers.

Takahisa Izumi is a geography teacher at Senshu University Matsudo High School and a lecturer in the Senshu University Teacher Training Course. His major research fields are geography education and education for social studies. He has written "Geography Education to Foster Global Citizenship: An Essay on Curriculum Development in Geography Education" (*Secondary Social Studies*, 27, 2008, in Japanese) and *Geography Teaching for Connecting Local Areas and the World* (2012, in Japanese), and related books and papers.

Hyunjin Kim is a lecturer in the Social Studies Department at Hokkaido University of Education. She was born on Jeju Island in South Korea. Kim received her master degree in education at Ewha Womans University in South Korea and her doctoral degree in education at the University of Tsukuba in Japan. Her research interests include international trends in geographical curricula and geographical inquiry-based learning. Recently, she has written "Geographical Inquiry-based Learning in Japan" (*Shin Chiri*, 57, 2009), and "International Trends in Geographical Education through Curriculum Analysis" (*E-journal GEO*, 7(1), 2012).

Tomoko Murayama is a professor of social studies in the Faculty of Education at Ibaraki University. After teaching in junior high school for several years, she became a University faculty member. Her major research fields are in social studies curricula and geography education focusing on the historical period since the Meiji era. Murayama has edited the geography textbook for junior high schools.

Shuichi Nakayama is an emeritus professor of Hiroshima University and Hiroshima University of Economics. He has taught geography and geography education since 1970 at several universities, mainly Hiroshima University. He served as a member of the Japanese National Commission for UNESCO from 2001 to 2007. His expertise includes the regional geography of South Asia, and geography curriculum development. He is the author of *Chiri ni mezameta amerika* (*America awakened to geography*) (1992, in Japanese). He is co-editor of *Jizoku kanou na shakai to chiri kyouiku no jissen* (*A sustainable society and the practice of geography education*) (2011, in Japanese). In addition, he has written many articles related to ESD and geography education.

Koji Ohnishi is an associate professor in the Department of Humanities at the University of Toyama. His major research focus is on children's geography education and behavioral geography. He has written "The important role of geography in making a crime prevention map for children" (*E-journal GEO*, 2, 2007, in Japanese), and related books and papers.

Hideki Oshima is a part-time instructor at Keio University, Tokiwa University, and Tokyo Tanashi Technical High School. He has worked as a construction consultant at Creative Research and Planning Co., Ltd. and is a special research associate at the Graduate School of Media and Governance at Keio University. His major research fields are geographic information science for geography and urban planning education. He has written "Teaching material development for satellite data interpretation" (*Map, the Journal of the Japan Cartographers Association*, 49(2) (2011, in Japanese), *The genealogy of geo media* (2010, in Japanese) and related books and papers. He operates a website in Japanese entitled "Easy digital maps" (https://sites.google.com/site/dejichizu/).

Takashi Shimura is a professor in the Social Studies Department at Joetsu University of Education. He previously taught high school for 17 years, and then joined the academic faculty at the University in 2002. His major research fields include geography education curricula, especially of the British Isles, and theories of map learning. He has written "Japanese Curriculum Reflections" in *Teaching Geography* Vol.34 No.2 (2009), *Recent Development in Geography Education in England* (2010, in Japanese), and related books and papers.

Hirokazu Takeuchi is a professor in the Social Studies Department at the Faculty of Education, Chiba University. He taught in the junior high school attached to Tokyo Gakugei University for 8 years, and then became a member of the University in 1989. His major research field is a demonstrative study about the formation of social awareness and citizenship. He has written "What children gain in social participation learning: A follow-up survey of children who participated in the 2001 public square revival project for the front of Shin-Urayasu station" (*Shin Chiri*, 58(1) (2010), and related books and papers.

Jun'ichiro Takata has been an associate professor in the School of Education at Gifu Shotoku Gakuen University since 2007. He taught geography at the high school level for 23 years, including 14 years at Hiroshima University Attached Junior and Senior High School. During this time, he taught geography teaching methodology at Hiroshima University for 7 years as a part-time lecturer. His major research fields are the study of developing instructional materials for social studies education; geography education, especially landscape studies related to regional awareness; and the study of UNESCO education. He has written a book entitled *Mou hitotsu no chiri*: *Another geography* (2000, in Japanese) and related papers.

Yumiko Takizawa is a professor of geography at Teikyo University. Takizawa worked for the Ministry of Education, Culture, Sports, Science and Technology as a senior specialist textbook examiner, and then became a member of the University in 2002. She primarily studies settlement geography and map education. She co-authored *Chimei no hassei to kinou*: *Nihonkai chimei no kenkyu* (*Origin and Functions of Geography Names*: *Study on the Japan Sea as a Geographical Name*) (2010).

Kiyoshi Teramoto is a social studies professor in the Department of Education at Tamagawa University. He taught at Tsukuba Elementary School, and then became a faculty member of the Aichi University of Education in 1984. His major research fields are geography education curricula, especially of the United States, and theories of cognitive maps. He has written the book *Map and Globe Skills for Teaching Social Studies* (2012).

Katsuki Toida was born in 1960. He is a professor in the Faculty of Applied Sociology at Kinki University. He taught high school for 12 years, and then became a faculty member of the University in 1996. His major research fields include human geography, folklore of Japan, and geography and history education. He has written "A Consideration of the Geography Curriculum for High School Education" in *E-journal GEO Vol. 7 No. 1 (2012)*, "A Subject Organization of High School Geography and History" (*Shin Chiri, 60(1), 2012*), as well as several high school geography textbooks.

Fumio Wada is a part-time lecturer of human geography and geography education at Fukuyama University and the Hiroshima University of Economics. He taught high school geography for 35 years, including 15 years at the Hiroshima University Attached Fukuyama Junior and Senior High School. His major research fields are regional geography and geography curriculum development. He is co-editor of *Jizoku kanou na shakai to chiri kyouiku no jissen* (*A sustainable society and the practice of geography education*) (2011, in Japanese). He has written many articles relating to geography curriculum development.

Kotaro Yamagata is professor in the Social Studies Department at the Joetsu University of Education. His major research field is physical geography, especially geomorphology and geoecology. He has written "Recent Grain-Size Coarsening of Floodplain Deposits and Forest Decline along the Kuiseb River, Namib Desert, Namibia" (*African Study Monographs*: *Supplement*, 40, 2010), as well as other related books and papers.

Kazuyoshi Yoshida is an associate professor in the Department of Education at Soka University. He taught elementary school for 30 years before becoming a University faculty member in 2012. He has studied children's environmental perception and written "Children's Play Behavior and Development Process of Environmental Perception" (*Geographical Review of Japan*, 81(8), 2008, in Japanese).

Hiroya Yoshimizu is a professor in the Curriculum and Instruction for Expert Teachers Course at Hyogo University of Teacher Education. He taught junior and senior high schools for 15 years, and then became a faculty member of the University in 2003. His major research fields include geography education curricula and lesson studies. He has written "Development of Multi-scale Geography Lessons Based on the Concept of Geographical Scale" (*Shin Chiri*, 59 (1), 2011), and related books and papers.

Minori Yuda is an assistant professor at the Center for Spatial Information Science at The University of Tokyo. She obtained her doctoral degree in the utilization of GIS in school education from Kanazawa University. Her research interests are the geography education of Japan and other countries, as well as course development to support students' spatial thinking skills in school education.

Introduction

The nature of geography education in Japan has rarely been communicated to the outside world. The goal of this book is to fill that gap and publicize Japan's geography education internationally by describing it systematically and in detail. In order to describe Japan's geography education system in detail, this book is made up of five different parts. Part I explains the current status of geography education within elementary schools, junior high schools, and senior high schools in the Japanese school system. Consideration is also given to detailing geography textbooks and the official screening system they must pass. In Part II the focus is on the learning content of Japanese geography education, and we try to elucidate the characteristics of Japanese geography education, arranging them into six separate themes: regional descriptive geography lessons, field research, maps-globes, and cognitive space, social participation, education for sustainable development (ESD), and disaster preparedness. In Part III, we discuss geography teacher training and licensure procedures required to teach geography education. We continue by explaining the assistance that geography academic associations give to geography teachers and the support for teacher adoption of geographic information systems (GIS). In Part IV, focus is given to geography education research which provides theoretical support for the practical implementations. We review current geography education research and their significance in Japan to offer an overview of what is known about geography learning and teaching in Asia. Based on the contents of Parts I through IV, Part V considers the future of geography education in Japan. The Great East Japan Earthquake and Tsunami Disaster in March, 2011 delivered a major shock to the Japanese geography education community. Thus, we consider the direction of Japan's geography education after having had this experience. Taking the current situation into account, we discuss the future of geography education as a part of Social Studies as well as the geography curriculum in senior high schools. Finally, we describe the characteristics of Japanese geography education with an eye towards the future, and suggest a path to promote the spread of geography education.

In this book we use the following terms in the interests of uniformity: The guidelines for instructional content that are legally mandated, the *gakushū shidō yōryō*, are sometimes translated as "the course of study" but in this volume it is translated as "the National Curriculum Standards." Subject names are capitalized. The Ministry of Education, Culture, Sports, Science, and Technology will be abbreviated as "MEXT". The following names will be abbreviated as: the Association of Japanese Geographers (AJG), AJG-Commission of Geographical Education (AJG-CGE), the International Geographical Union Commission on Geographical Education (IGU-CGE).

Yoshiyasu Ida

Part I
Overview of Geography Education in Japan

Chapter 1
Outline of Geography Education in Japan

Yoshiyasu Ida and Takashi Shimura

Abstract The aim of this chapter is to present an overview of the current situation of geography education in Japanese schools and consider its future in the country. Geography is part of the Social Studies curriculum in elementary and junior high schools, and it is an elective subject in senior high school. The main content of the geography curriculum in elementary school concentrates on the students' local environment. Japanese and world geography is the focus in junior high school, whereas world geography comprises the main content in senior high school. As for teacher training, there is an in-service training system in Japan. Furthermore, a teacher's license update system was recently implemented requiring compulsory training, which in some universities includes fieldwork training. This chapter also addresses political issues affecting geography education and surveys how geography education groups can appeal to the government and the public, to highlight the importance of geography in schools.

Keywords Geography education • In-service teacher training system • Japan • New National Curriculum Standards

1.1 Education System and Geography

1.1.1 Overview of the Education System and Geography-Related Subjects

The latest version of the National Curriculum Standards in Japan was published in 2008/2009. While the former curriculum emphasized methods of learning, the new curriculum focuses on the importance of acquiring fundamental knowledge. This revision is grounded in the philosophy "Bringing Knowledge Back In" (Young 2008) that is seen in developed countries in the 21st global competitive society. Geography education is not an exception to this trend.

Y. Ida (✉)
University of Tsukuba, 1-1-1 Tennodai, Tsukuba, Ibaraki 305-8572, Japan
e-mail: ida@human.tsukuba.ac.jp

T. Shimura
Joetsu University of Education, Niigata, Japan

Y. Ida et al. (eds.), *Geography Education in Japan*, International Perspectives in Geography 3, DOI 10.1007/978-4-431-54953-6_1

We first present an overview of the current situation of geography education in Japanese schools and consider future teaching options available, with a focus on curricula and in-service teacher training.

Since the beginning of modern education in Japan, in the late nineteenth century, the school system has been strictly administrated by the Ministry of Education (MEXT).[1] The government established a uniform national education system by means of the Educational Ordinance in 1872 and Education Order in 1879, which included the National Curriculum Standards. This marked the first educational reform in Japan. In this curriculum system, Geography was an independent and compulsory subject, in addition to History (Japanese history) and Morals (*Shyuushin*) in elementary schools.

In 1947, just after World War II and under the Allied occupation, there was a second educational reform that used the U.S. educational system as its reference. The school system adopted a 6-3-3 structure (6 years of elementary school, and 3 years each of junior high and senior high school). The length of compulsory education was increased from 6 to 9 years, extending the school age of students from 6 to 15 years old. This system has continued to the present day; elementary and junior high schools remain compulsory.[2]

Under this education system, MEXT enacts and revises the National Curriculum Standards at approximately 10-year intervals. The first National Curriculum Standards reform in 1947 introduced the new subject area of Social Studies borrowed from the U.S. curriculum. As a result, geography and history were integrated into Social Studies which emphasized problem-solving skills. However, a systematic method of learning was adopted for Social Studies after 1955. This measure was taken for various reasons, including the fact that the ability to solve problems only provided fragments of knowledge.

Learning centered on problem-solving was criticized, and a systematic method of learning was feared that this instructional method stressed cramming of information or overly emphasized rote learning. This led to careful selection of learning content and experiences for the 1989 National Curriculum Standards. In elementary schools, Social Studies and Science were abolished in the first and second grades and replaced by Life Environmental Studies, which emphasized experience. This change marked the beginning of the third educational reform, which is used in the present education system. In the 1989 revision, high school level Social Studies was separated into two new subjects: Geography and history, and Civics.

In recent years, new criticism arose faulting the current school curriculum for failing to teach basic knowledge. In response, the revised 2008/2009 National Curriculum Standards once again emphazed the acquisition of basic knowledge in all subjects. The new curriculum also focuses on social participation, or means of personal contribution to society. Table 1.1 outlines the current Japanese school system and geography-related subjects.

---

[1] The current office is the Ministry of Education, Culture, Sports, Science and Technology (MEXT).

[2] All senior high schools have an entrance examination. However, the percentage of students who attend high school is currently around 98 %.

Table 1.1 School system and geography-related subjects in Japan

Age	School year (grade)		School type	Compulsory subject		Elective subject in Geography and History		Elective subject in Civics	
17–18	3		Senior High school	Geography and History	World History A/B	(at least one required)	Geography A/B	(at least one required)	Contemporary society
16–17	2						Japanese History A/B		Ethics
15–16	1								Politics and Economy
14–15	3	Compulsory education	Junior high school	Social Studies	Civics field				
13–14	2				Geography field			History field	
12–13	1								
11–12	6		Elementary school	Social Studies					
10–11	5								
9–10	4								
8–9	3								
7–8	2			Life Environment Studies					
6–7	1								

1.1.2 Geography in the Japanese Education System

1.1.2.1 Social Studies Curriculum in Elementary Schools

In the first National Curriculum Standards (1947), Social Studies was considered the most important subject, known as a core subject, because of the significant role it played in creating a democratic society in Japan. Therefore, teaching Social Studies involved many hours of lessons, comprising 16.5 % of total elementary school classroom hours. Since then, however, the number of hours dedicated to Social Studies have decreased.

The implementation of a new subject area, Life Environmental Studies, replaced Social Studies and Science at the first and second grade levels in 1989. By 2002, a new subject called Integrated Studies, included in all grade levels, had a huge impact on Social Studies education. After 2002, only 6.4 % of classroom hours were dedicated to Social Studies (Tabe 2004). Table 1.2 shows unit hours allocated to Social Studies per year (MEXT 2008). The curriculum regulation defines one school year as having 35 weeks of instruction, which translates to only 2 or 3 class hours per week of Social Studies from third to sixth grade.

Overall, the aims of elementary Social Studies in the National Curriculum Standards are[3]: (1) to develop an understanding of life in society; (2) to develop an understanding and love for Japan's land and history; and (3) to foster foundational qualities of citizens who are members of a peaceful and democratic country and society. These aims respectively correspond to an understanding of Japanese society, geography, history, skills, values, and attitudes. This structure is the same in each school grade, while the final goal of Social Studies is to foster students' understanding of basic contributions as a citizen to the nation.

Social Studies in elementary schools encompass broad and comprehensive content. The term "Geography" is not used in textbooks, but many geography-related topics are included in all grades although the content and expectations are different across each grade level (Table 1.3). In this table, columns are divided by strands into geography, history, and civics. The instruction emphasis is on geography and civics in grades 3 and 4, geography in grade 5, and history and civics in grade 6. However,

Table 1.2 Unit hours allocated to Social Studies per year (National Curriculum Standards 2008 edition)

	Elementary				Junior high school		
Grade	3rd	4th	5th	6th	1st	2nd	3rd
Social Studies	70	90	100	105	105	105	140
Total unit hours of schooling	945	980	980	980	1,015	1,015	1,015

Course length: 1 unit hour is 45 min in elementary school, 50 min in junior high school

[3] There is no English version of the course of study in Social Studies that is authorized by MEXT. The text is a translation by the authors.

Table 1.3 Social Studies curriculum (2008 edition) in Japanese elementary schools

	Geographical strand				
Grade	Place	Theme	Skill	Historical strand	Civic strand
3	Local area (city, town, village)	Landforms, land use, and traffic	Observation, investigation, blank map, thinking about feature differences by places		
		Commercial activities of people in the local community and their relationship with other areas in Japan			Public utilities
					Electricity, gas, and water supply, and waste management
4			(Atlas)	Lifestyle changes and the achievements of ancestors	Activities to keep safe from disasters and accidents
	Prefecture	Characteristics of the prefecture and its relationship with other areas in Japan and foreign countries	Thinking about the characteristics of the prefecture		
5	Country (Japan)	Land of Japan and the relationship between the physical environment and people's lifestyle and industries	Map, globe, various kinds of materials, visits		
		Agriculture and fishery in Japan, their role in food supply and relationship with the physical environment			Agriculture and fishery in Japan, their role in food supply and relationship with the physical environment
		Manufacturing industries in Japan and their role in our lives			Manufacturing industries in Japan and their role in our lives
		Communication industries in Japan and their influence on our lives			Communication industries in Japan and their influence on our lives
6			Map	Main historical matters seen through the achievements of ancestors and cultural heritage	
					Constitution and function of Japanese politics
		Place of Japan in the world			Place of Japan in the world

it should be noted that this division is not absolute, students also learn about different industries in Japan in grade 5.

In Table 1.3, the geography strand is divided into place, theme, and skill. The concept of "place" follows a progressive expansion from the student's local community in grade 3, to their prefecture in grade 4, and then to the entire country in grades 5 and 6. This is based on the expanding horizon (environment) curriculum theory or the widening horizon curriculum theory, which is generally adapted in the development of the Japanese Social Studies curriculum.

Skills are prescribed in the most detail in grades 3 and 4, however, in these grades, it is too late for students to begin learning about maps. When Social Studies was included in grades 1 and 2, foundational map skills were taught in these grades. But, the abolition of Social Studies in the 1989 National Curriculum Standards revision shifted the introduction of mapping skills until grade 3 and the use of a globe is not taught until grade 5. These curriculum sequences are inappropriate for students' development stages.

In Japan, school textbooks and atlases authorized by MEXT must be used in classrooms. The textbooks are written and edited by private-sector publishers, and later approved by MEXT in accord with the National Curriculum Standards. Textbooks and atlases are distributed to all elementary school students free of charge. Students receive an atlas in grade 4, a year after Social Studies instruction begins, and use it throughout elementary school. Some geography educators and researchers suggest that atlases should be distributed to students prior to grade 4 to develop their curiosity about the world. Also, students should not use the same atlas for 3 years but instead receive updated copies every year.

1.1.2.2 Geography Curriculum in Junior High Schools

In junior high schools, Social Studies includes the subjects Geography, History, and Civics. Geography and History are usually taught as independent courses within a school timetable, each with a different school textbook.

Traditionally, the conventional geography curriculum in Japan was based on regional geography. However, in 1998, a significant National Curriculum Standards reform adopted a "sample studies" approach. This approach is similar to what is called "case studies" in the British geographical education, which involves choosing situations or places from around the world to illustrate geographical theories or themes, as opposed to focusing on regional characteristics (Shimura 2009a). This reform meant abandoning the traditional regional geography framework, but it was unpopular and disliked by schoolteachers, educational researchers, and society.

The most recent 2008/2009 curriculum edition returns to a focus on regional geography, with many teachers welcoming this reform. The overall aim of junior high school Social Studies is similar to that of elementary schools. The objectives of geography are to develop an understanding of: (1) the land of Japan and regions of the world; (2) the regional characteristics and issues seen through geographical perspectives and thinking; (3) the interdependence of regions, the differences and similarities of regional characteristics, and the different regions; and (4) to foster the skills and attitudes that enable students to consider, resolve, and represent geographical matters through various activities including fieldwork.

The 2008/2009 version of the National Curriculum Standards consist of two main parts (Fig. 1.1), namely world regions and Japanese regions, with both parts

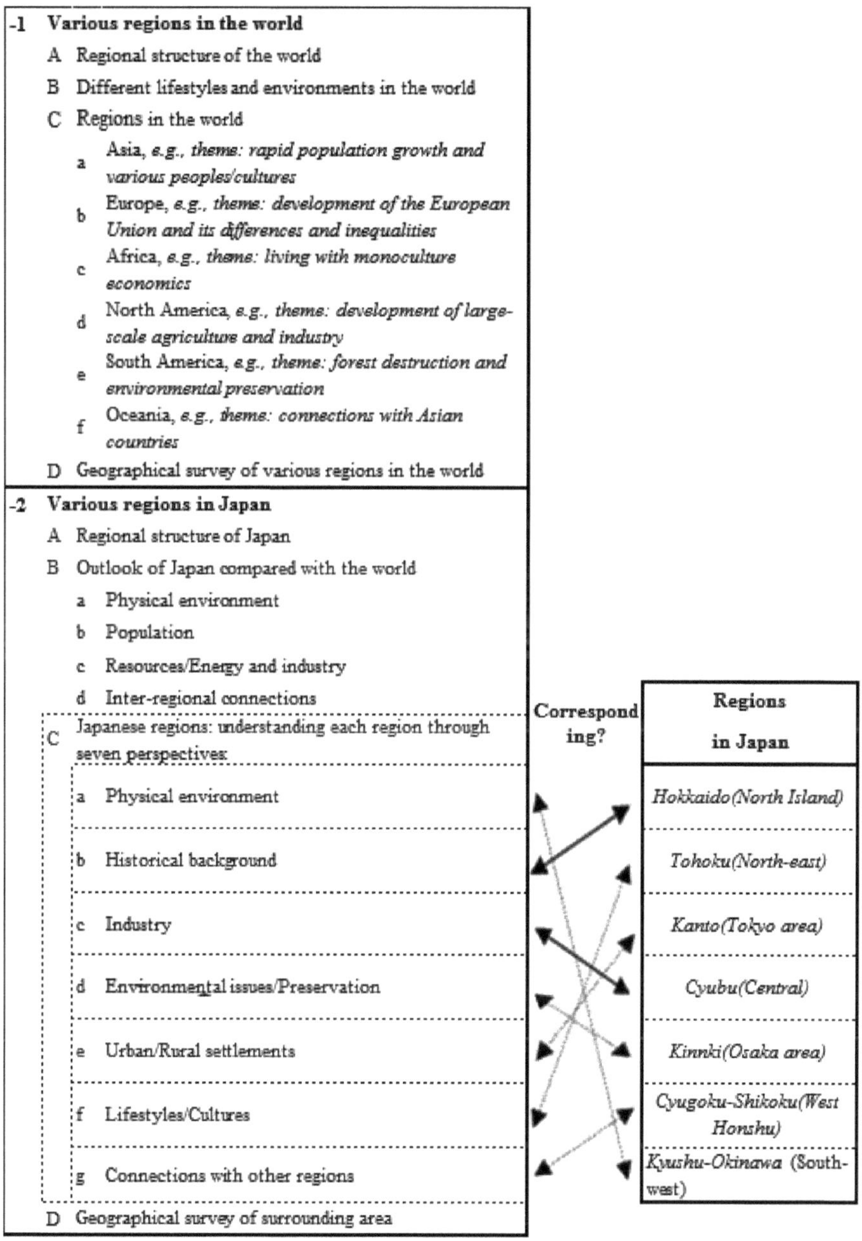

Fig. 1.1 Geography curriculum (2008 edition) for Japanese junior high schools. *Source*: Shimura (2009b)

requiring regional coverage. This reform requires an innovative strategy for teaching regional geography, by focusing on a theme for each region. Exhaustive teaching of the subject is cautioned against. This strategy is challenging and contrasts sharply with the previous version (Shimura 2009b). Since 2012, junior high geography lessons cover the entire world through six continent or state divisions, and the whole of Japan through the traditional seven regional divisions.

1.1.2.3 Geography Curriculum in Senior High Schools

In 1978, a new subject entitled "Contemporary Society" was introduced to the Social Studies section of the National Curriculum Standards, and this subject was compulsory from the first grade of senior high school. As a result, the number of students enrolled in Geography decreased significantly, because they could choose Contemporary Society as an alternative subject.

Since the 1989 curriculum revision in the third educational reform, geography has been a subject in the Geography and history area, as stated above. This subject area consists of six different divisions: Geography A (two credits/two unit hours per week), Geography B (four credits/four unit hours per week), Japanese History A (two credits), Japanese History B (four credits), World History A (two credits), and World History B (four credits). World History is compulsory, whereas Japanese History and Geography are elective; students are required to obtain at least four credits total in the subject area. Because of this requirement, only an estimated 50 % of all senior high school students are taught Geography.

In March 2009, MEXT revised and enacted a new senior high school curriculum to be implemented in 2013. The overall goals of Geography and history area are to develop an understanding of: (1) the historical course in Japan and the world; (2) the regional characteristics of people and cultures in Japan and the world; (3) to foster awareness and the qualities necessary for living independently in a democratic and peaceful state or society internationally. The first and second aims demonstrate the historical and geographical goals, respectively, with major emphasis given to developing understanding and knowledge. The third aim, however, suggests the ultimate goal of studying geography and history, with the main emphasis on values and attitudes. This final aspect has been included as the primary objective of Social Studies.

We observe that the significance of values and attitudes is less emphasized in current Geography and History courses compared with past Social Studies subjects in senior high schools. This suggests that Geography and History curricula have become more oriented toward developing understanding and knowledge.

Tables 1.4 and 1.5 below illustrate the contents of Geography A/B in the 2009 curriculum. The previous Geography A curriculum was based on issues and skills. Although the new curriculum develops these ideas, it focuses more on local areas, as in part (2) of Table 1.4. A prime example is the inclusion of disaster prevention, a topic emphasized in the geography curriculum. Geography A has an independent unit entitled "Natural environment and disaster prevention," and this theme is also

Table 1.4 Geography A curriculum (2009) in Japanese senior high schools

(1)	Geographical considerations of current global characteristics and issues
	A. Representation of the modern world through globes and maps
	B. Diversity of lives and cultures in the world
	C. Geographical considerations of global issues
(2)	Geographical considerations of local issues
	A. Maps in everyday life
	B. Natural environment and disaster prevention
	C. Geographical issues in the local area and fieldwork

Table 1.5 Geography B curriculum (2009) in Japanese high schools

(1)	Maps and geographical skills
	A. Geographical information and skills
	B. Applications of maps and fieldwork
(2)	Systematic geography considerations in the current world
	A. Natural environment
	B. Resources and industries
	C. Population and settlements
	D. Cultures and people/religion
(3)	Regional geography considerations in the current world
	A. Regional divisions in the current world
	B. Regions in the current world
	C. Japan in the current world

taught in Geography B and in the junior high school curriculum. As a result, school-level geography is expected to make a critical contribution to disaster prevention education.

The Geography B curriculum has changed dramatically (Table 1.5) from the previous version (1999). The last version emphasized skills and geographical ways of thinking. Students thus learnt how to survey regions on different scales, but did not focus on understanding regional characteristics. Students were required to study only two or three regions at each scale, i.e., community, national, and continental (Shimura 2010). This differed from the traditional regional geography curriculum approach that focused on regional characteristics. However, like the junior high school reform, this innovative curriculum was not welcomed in the educational field.

The revised 2009 Geography B curriculum consists of three main parts, with a theory of development differing from the 1999 version. The three parts comprise skills, a systematic, and a regional geographical approach. The third element includes subsection B, "Regions in the current world," in which students learn about the various regions without spatial bias while applying various regional geography learning strategies. These strategies include static, dynamic, or topical, and comparative regional geography. Instruction in regional geography differs from traditional regional geography strategies.

1.2 In-service Training Programs for Geography Teachers

1.2.1 In-service Training Programs

There is a wide variety of in-service training programs for teachers in Japan. All teachers, regardless of the subject matter, must undergo an initial inservice training program after successful recruitment. Over a one-year period, teachers are required to undertake both in-school (10 or more hours per week, 300 h per year) and out-of-school (25 days per year) training, in addition to the regular teaching load at their school.

After completing the initial in-service training, teachers continue their professional development throughout their career. According to Murata and Yamaguchi (2010), in-service training for teachers is essentially divided into three types: self-training, in-school, and in-service training provided by various organizations. The latter is training provided by educational administrative bodies and in-service training provided by universities. Since this chapter concerns professional development for geography teachers, we discuss the in-service training programs furnished by geography-related academic societies, as well as those that form part of the teacher's license renewal system. Since the implementation of the system in 2009, a teaching license is renewable every 10 years.

1.2.2 In-service Training Programs Provided by the Association of Japanese Geographers

The largest professional geographical society in Japan is the Association of Japanese Geographers (AJG), with approximately 3,000 members. The Commission of Geographical Education (CGE) is an AJG committee tasked with the mission to disseminate geography education. The AJG-CGE provides in-service teacher training support to promote geography in the educational system which targets elementary and junior high school teachers, who in general are not especially well versed in geography. These workshops are intended to improve teachers' general knowledge of geography. Similar professional development opportunities are offered by other geographic societies, such as the Human Geographical Society of Japan and universities via citizen seminars.

In Japan, research associations exist for educators of different subjects, which are supported by prefectural and municipal boards of education (government) and national-level research associations for geography teachers. Each association conducts activities centered on their own programs. Currently, there is a gap between academic societies that primarily serve researchers and educational organizations whose members are practitioners from elementary, junior high, and senior high schools. The inherent differences in the purposes of these two groups set them apart. Despite these differences, there have been efforts to work together, with examples

Table 1.6 Content of in-service training for teachers provided by the AJG-CGE

Subject	Content
Interpretation of topographical maps and work using a literary work (1.5 h)	• Trainees focus on the route used by the main character in a literary work to commute to work by bicycle; trainees read the text and verify it against a topographical map.
	• Trainees realize that the roads mentioned are not found on the topographical map. As they used a modern topographical map, they recheck the roads using a topographical map from the time of publication (1920s). They therefore learn the usefulness of old maps and their comparison with up-to-date maps.
	• Trainees imagine the landscape of the area from symbols on the topographical map, and compare the landscape in the film version of the literary work with the imagined landscape.
Use of maps with a computer (2.0 h)	• Overview of map analysis methodology
	• Trainees collect statistics on the Internet, and prepare a statistical map using the statistical information collected.

of academic societies proposing plans for professional associations to partner teacher in-service training projects.

However, in Japan, there is talk about a "crisis of geography" in school education. Since the official announcement of the National Curriculum Standards in 1989, Geography has been excluded as a compulsory subject at many senior high schools. As a consequence, the need has arisen for geography researchers and educational practitioners to establish a cooperative structure aimed at promoting the subject. This is one reason why the CGE was established by the AJG in 1998, and why some in-service training programs provided by the AJG-CGE are incorporated into training workshops conducted at the prefectural or municipal level.

The AJG-CGE provides several types of training programs. Table 1.6 provides examples of the most frequent training programs.

An elementary school teacher who participated in a training workshop pointed out that the course entitled "Interpretation of topographical maps and work using a literary work" improved her reading comprehension, but that it could also be used as teaching material in Japanese. In Japanese schools, the study of topographical maps often does not extend beyond memorizing map symbols and reading contours. This approach is thought to effect a child's interest in the subject, leading them to be "allergic" or averse to maps. The in-service training aims to revive children's interest and is directed at supporting a wider use of topographical maps, for example, when reading various literary works. This form of training is effective in the sense of familiarizing trainees with those maps.

The course "Use of maps with a computer" includes content that may be considered introductory to Geographic Information System (GIS). Geography teachers in elementary and junior high schools seldom have an opportunity to collect statistical information on the Internet and create maps on computers; therefore, this type of training helps them promote the use of computers in geography. In other words, the course will lead to the diffusion of GIS.

The exemplary training workshops introduced described above require about 4 h although the duration can be adjusted depending on the situation. Elementary and junior high school teachers are not the only targets for training, but also high school teachers of geography and history.

1.2.3 Renewal System for Teacher's Licenses

In the elementary school system, a variety of subjects are taught by one teacher, however, they may not be fully prepared or trained to teach all the subjects. Even if the content or learning methods in Social Studies or geography are excellent, some teachers may need help to use them effectively for instruction. This is not only applicable to Social Studies, but to all subjects. In response to this situation, MEXT introduced a renewal system for teacher's licenses in 2009 so that they keep up to date on the latest educational developments. The system requires elementary, junior high, and senior high school teachers to receive training every 10 years and to pass examinations in order to have their teacher's license renewed. Teachers who are around the ages of 32, 42, and 52 years are required to participate in in-service training for their license renewal. The training takes place at universities and similar institutions, led by university lecturers.

As established by the government, in-service training in the license renewal system requires at least 12 h to learn about the latest educational developments, and at least 18 h for course instruction and student guidance. At Tsukuba University, one of the universities that most actively conducts in-service training for the license renewal system, lecturers who specialize in pedagogy primarily teach the latest educational developments as a compulsory subject. Lecturers who are experts in course instruction and student guidance offer a variety of lectures as elective subjects. During the in-service training program conducted in 2010 for the teacher's license renewal system, Tsukuba University offered a total of 77 elective subjects, including a geography-related subject entitled "Lesson structure in Social Studies (Geography and history) with field research" (6 h). The background of this subject is as follows.

In Japanese elementary and junior high schools, Social Studies focuses on learning about the local area of the students. For this reason, field research is an important learning outcome in Social Studies. In 2006, 81 students from Tsukuba University took a survey that asked whether they undertook field research in Social Studies-related lessons. The responses showed that 74 % of students conducted fieldwork in elementary school, 12 % in junior high school, and 4 % in senior high school. Although the rate of fieldwork experience in elementary school was relatively high, the overall rate, including high schools, was rather low. One of the most important factors influencing fieldwork is that teachers, including those in elementary schools, do not really understand how to plan and conduct field research and what outcomes to expect. For this reason, it was considered necessary to offer lectures for teachers to learn about the significance and methods of fieldwork and how to verify its effects.

Table 1.7 Lecture content of "Lesson structure in Social Studies (geography and history) with field research"

Schedule	Outline
1. Overview	Lecture on the definition, usage, significance, and necessity of field research.
2. Reading topographical maps of the university area	Identify areas around the university that are considered geographically characteristic by reading a topographical map (the class is divided into groups for discussion).
3. Creation of a route map	Create a route to follow by incorporating walking spots that are considered to be geographically representative.
4. Field observation (about 2 h)	Follow the route map on foot (the lecturer leads the trainees and explains the geographic events as needed).
5. Summary after observation	Return to the classroom. Compare observations with the literature and an old version of the map, give appropriate explanations, and respond to any questions.
6. Visit to the Geographical Survey Institute	Visit the Geographical Survey Institute, which has a vast stock of maps, attend a lecture on the creation of maps, and tour the facilities.

The fieldwork professional development program accepted 40 trainees. Elementary, junior high, and senior high school teachers each accounted for approximately one-third of participants attending this lecture. Table 1.7 shows the content of the lesson

After completing this workshop, a few of the trainees who normally incorporated field research into their lessons reported increased confidence, while those who rarely conducted field research or did not know how to use it in their lessons realized the importance of actually visiting a site. They also learned methods necessary to produce teaching materials based on the area surrounding their schools. Teaching field research methods as part of the in-service teacher training for license renewal is extremely effective for the dissemination of geography as a subject. However, decisions about the content of such training programs are left to each university. Although some universities provide in-service training workshops to promote the understanding of geography, these are in the minority.

As seen above, although there are in-service training programs that provide high-level skills and advanced knowledge to geography teachers, the recent trend in geography education involves measures to raise the teaching skill of geography teachers, especially in elementary and junior high schools.

1.3 Problems and Future Directions in Geography Education

1.3.1 Current Problems in Geography Education

The Japanese educational system emphasized "learning how to learn" during the 1980s and 1990s. However, in the 2000s, education content shifted to a thorough acquisition of "basic and fundamental" knowledge. This trend was also visible in

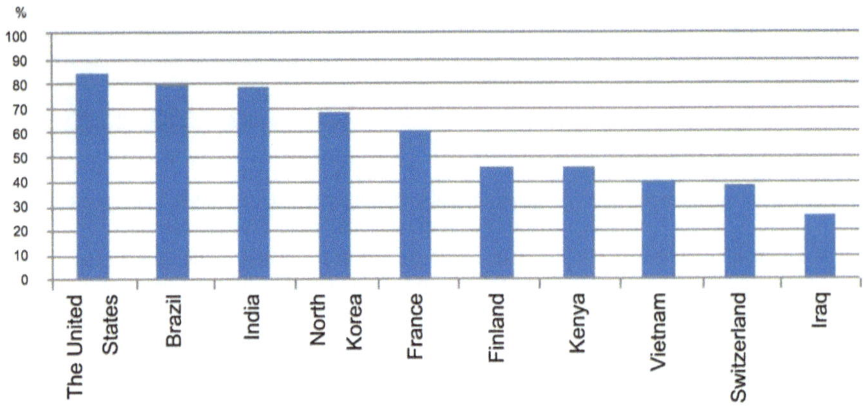

Fig. 1.2 Percentage of Japanese students who correctly located countries on a world map. (Based on AJG-CGE 2008). *Note*: The survey was conducted by the AJG-CGE between December 2007 and February 2008 on 6,159 students from 51 high schools throughout Japan

geography education. In addition, increasing numbers of students are electing not to take geography courses in senior high school. As a result, there are many students who are unable to identify countries on a world atlas, as shown in Fig. 1.2.

In light of the lack of geographic knowledge, three problems have been identified in Japan's geography education. The first concerns learning content, which is based on the systematization of geography, including topography and physical geography. Instruction in geography content based on themes is possible, and it is necessary to include appropriate content for learners to develop into responsible citizens. The second problem is the development of skills. The National Curriculum Standards set forth by MEXT states that skills such as map reading and geographical thinking should be emphasized. However, the National Curriculum Standards does not refer to the progressive development of skills, as highlighted by Ida (2008a, 2008b, 2008c). An important issue for the future of geography education in Japan is therefore to determine which skills must be taught and how they should be developed, as students' progress through their schooling.

The third problem is the exclusion of geography from compulsory subjects in senior high school. As mentioned above, Geography, World history, and Japanese history constitute the single elective subject area of Geography and history, but only World history is compulsory. Although some prefectures require senior high schools to teach Japanese history, only an estimated 50 % of high school students take a geography course. It is thus vital for groups involved in geography education to appeal to the national government and society at large to highlight its importance, such that the current National Curriculum Standards will be revised to ensure geography is compulsory for all senior high school students.

1.3.2 Future Directions in Geography Education

Geography an indispensable subject in education that contributes to the formation of character, but since it is not compulsory in schools, its usefulness is no longer clearly recognized. If enough people recognize the importance of geography education, its future will be guaranteed. To this end, groups that are actively engaged in geography research and education must continue to stress the importance of the subject to the national government and society at large. These groups should also present specific models for learning content and methods for skill development. If these groups fail to do so, geography education in schools will decline, which will have an adverse impact on the character development of school-age children. A bright future for geography entails a similar future for children. Today, geography education in Japan stands at a critical juncture.

References

AJG-CGE (2008) Daigakusei, Koukousei no chiriteki-ninshiki no Chousa houkoku (A survey report on the understanding of geography among university and senior high school students). http://www.ajg.or.jp/chirikyouiku20080319.pdf. Accessed 8 Dec 2013

Ida Y (2008a) Seakaitiri wo manabu tiri no wakugumi ha dou kawatte ikunoka, tiri o donoyoni apiiru surunoka (How the framework of geography with world geography will change? How should geography be highlighted?). Chiri Geppo (Month Rep Geogr) 505:1–2

Ida Y (2008b) Gakusyushidoyoryo no kanzen koryaku koko chiri rekishi komin (Total conquest of the curriculum guidelines, Senior high school geography, history, and civics). Kyoshoku Katei (Teacher Train Course) 34(11):44

Ida Y (2008c) Gakunenjun ni shimesareta noryoku ni kansuru mokuhyo: fumikomenai dankaiteki sukiru no kakutoku (Goals of abilities for each grade—difficulty of phased development of skills). Shakaika Kyoiku 596:14–15

MEXT (2008) Chugakko gakushu shido yoryo (National curriculum standards for junior high school). Higashiyama shobo, Kyoto

Murata Y, Yamaguchi M (eds) (2010) Gendai nihon no kyoiku: seido to naiyo: bairingaru tekisuto (Education in contemporary Japan: system and content: a bilingual text). Toshindo, Tokyo

Shimura T (2009a) Bibliographical consideration of 'sample studies' and 'case studies' in England: from a viewpoint of Japanese geography education. Shin Chiri (New Geogr) 57:256–262

Shimura T (2009b) Japanese curriculum reflections. Teach Geogr 34(2):61–63

Shimura T (2010) High school geography in the 2009 Japanese national curriculum: reflections on Japanese social studies education history. Chiri kyoiku ESD foramu: denshi ban (Geographical ESD forum: web edition). http://www.esd-geography.com/wp-content/uploads/2010/03/71323 5ff6cb996006ebe5d5d83053bed.pdf. Accessed 19 Mar 2013

Tabe T (2004) Nyumonki ni okeru jiogurafikaru sukiru no nichibei hikaku to kadai (The Japanese-American comparison of geographical skills in the initiation term). Chiri kagaku (Geogr Sci) 59:132–139

Young M (2008) Bringing knowledge back In: From social constructivism to social realism in the sociology of education. Routledge, London

Chapter 2
Problems and Perspectives of Geography Education in Japanese Elementary Schools

Kazuyoshi Yoshida

Abstract This paper illustrates geography education in Japanese elementary school and investigates the current problems and perspectives at these grade levels. The problems are summarized as follows: (1) Students have few map skills and only basic knowledge of various place names in Japan and in the world and (2) students learn little about world geography through the elementary school curriculum. As a result, elementary school students have few opportunities to study physical geographic phenomena and to learn about features of the natural environment.

Keywords Elementary school • Geography curriculum • Map skills

2.1 Introduction

This paper investigates current problems in and perspectives on elementary school geography education in Japan. Geography is mainly taught in the form of two integrated subjects at elementary schools: Life Environmental Studies and Social Studies. The former is devised for students in the first and second grades while the latter is design for the third to sixth grades. Instruction is provided in accordance with the so-called expanding environments curriculum, which is based on a concentric circle type of study theory. Neighboring areas are taught first, and then towns and cities all over Japan (Fig. 2.1). Areas near the school are studied in lower grades, and more distant areas in higher grades. In the sixth grade, students start to learn about the geography of foreign countries. In this curriculum, greater significance is given to developing desired citizenship abilities for the discussion of contemporary issues, rather than to acquiring geographical skill and knowledge. Moreover, physical geography is nearly absent from the curriculum.

Geographers have commented on problems in Japan's elementary school level geography education (Nakayama et al. 1990). Therefore, it became necessary to reform the curriculum to solve the following problems:

K. Yoshida (✉)
Soka University, 1-236 Tangi-cho, Hachioji, Tokyo 192-8577, Japan
e-mail: ykazu@soka.ac.jp

Y. Ida et al. (eds.), *Geography Education in Japan*, International Perspectives in Geography 3, DOI 10.1007/978-4-431-54953-6_2

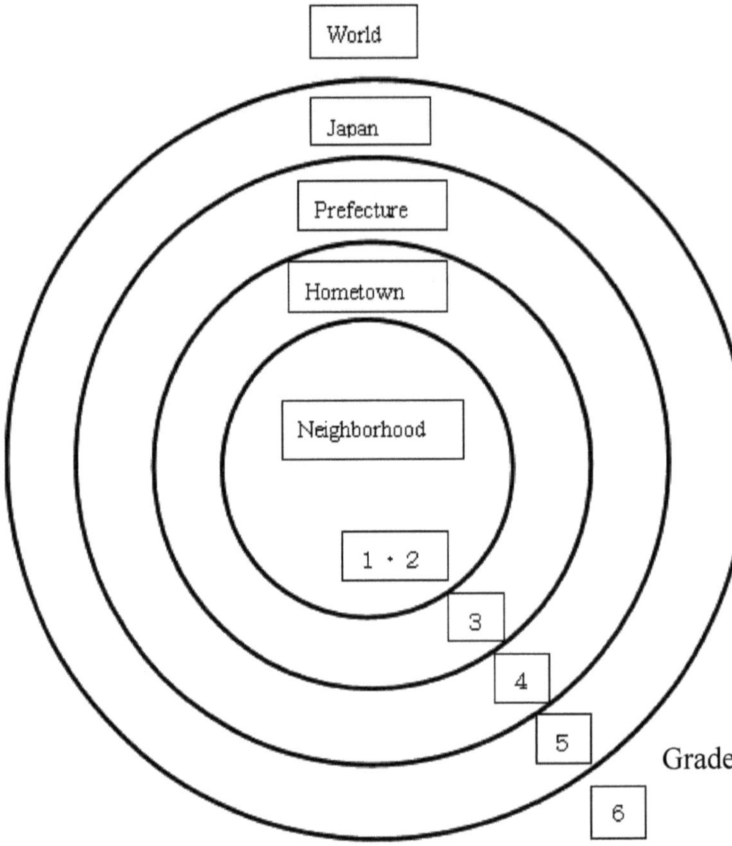

Fig. 2.1 The expanding environments curriculum

(1) Students have few map skills and only basic knowledge of various place names in Japan and in the world.
(2) Students learn little about world geography through the elementary school curriculum.

2.2 Geography Education in Elementary School Across Grades

The curriculum for elementary schools was recently changed. A new National Curriculum Standards for Social Studies was presented in 2008 by MEXT; the aims and content, including for geography education, were partly revised.

In the new curriculum, the practical use of a map and globe was recommended, as was the learning of basic knowledge (Table 2.1). Students from the third to sixth

Table 2.1 Main syllabus for Social Studies

Grade	Name and content of unit	Periods per year
3	Unit 1—Town where we live: neighborhood map making	70
	Unit 2—People at work in our town: supermarket, farming, factory	
	Unit 3—Old life in our town: traditional ways of life, old buildings	
4	Unit 4—Safety: police office, fire station	90
	Unit 5—Health: garbage disposal, recycling	
	Unit 6—Local development: irrigation waterway	
	Unit 7—Prefecture where we live: landform, land use	
5	Unit 1—Japan in the world: continents, oceans	100
	Unit 2—Food production: rice crops, fishery	
	Unit 3—Heavy industry and industrial areas: car manufacturing, import and export	
	Unit 4—Information industry: broad casting station, computers and the Internet	
	Unit 5—Land use and Japan's environment: protecting the environment, natural hazards	
6	Unit 1—Japanese history	105
	Unit 2—Japan's political system: the Diet, the Constitution	
	Unit 3—Countries of the world: U.S., China, Korea, etc.	

grades now learn the names and positions of all 47 Japanese prefectures. Additionally, in fifth grade they learn the names and locations of six continents using world maps: Eurasia, Africa, North America, South America, Australia, and Antarctica. They also learn about major oceans, namely the Pacific, Atlantic, and Indian. Thus, accurate map skills and rich locational knowledge help students develop their geographic understanding.

2.2.1 The First (6–7 Years Old) and Second (7–8 Years Old) Grades

During the first 2 years of elementary schooling, instruction on Life Environmental Studies is provided three times a week. A single period lasts 45 min. Attention is paid to personally observing geographical matters in neighboring areas.

Students learn about the location and use of playgrounds in parks near their schools. Sometimes, they observe flowers and trees during individual seasons and note changes in the atmospheric temperature and physical environment. Students learn about the location and function of stores, police stations, and play parks, through fieldwork and observation around the school. They sometimes use illustrated maps marking neighborhood locations such as stores, stations, bus stops, and post offices.

2.2.2 *The Third Grade (8–9 Years Old)*

Students usually learn to read and draw maps of the neighboring area at the beginning of third grade, that is, during the spring term. They begin by laying out pictures of buildings, trees, shops, and other objects on large sheets of paper spread out on their desks (Fig. 2.2). These activities allow them to consolidate their knowledge of their own neighborhoods. They learn about the spatial arrangement of places in their surrounding environment by making illustrated maps. Thus, students can develop their environmental perception and cognitive abilities (Yoshida 2008).

Eventually, study areas are extended to the town where the students live. They study landforms, land use, road and railroad patterns, shopping street, and factory locations. They compare land use and landscapes to other parts of town with a different physical environment. They read not only illustrated maps but also other general maps. Vertical aerial photographs can be used to help students read maps.[1]

2.2.3 *The Fourth Grade (9–10 Years Old)*

Students conduct fieldwork to observe firsthand the location and function of a fire station or police office. Before leaving their classrooms, they verify on a map the route from the school to their destination.

Fig. 2.2 Neighborhood map made by students

[1] The National Curriculum Standards' objectives for the third and fourth grades are combined. However, a separate syllabus is planned for each grade.

Study areas are extended to the prefecture to which the students' town belongs. For example, in Tokyo schools, students usually study landforms and land use, expansion of dormitory towns, and changes of land use in the Tokyo Metropolitan Area. They learn how farmland on plateaus in the western part of Tokyo was cultivated after construction of an irrigation waterway in the seventeenth century. Also studied are typical ways of life in towns, such as those in the mountains or on islands that depend on unique physical environments. Students are also expected to have the ability to read and use maps with scales and contour lines.

2.2.4 The Fifth Grade (10–11 Years Old)

Students learn about Japan's physical geography and its location in the world. They are introduced to economic geography while learning about agriculture, including rice cultivation, fisheries, land use, climate, and the distribution of factories in terms of food, machinery, metals, fiber, and chemical industries. Lessons also focus on car manufacturers that use various types of robots to make high quality goods. In addition, they learn that many Japanese companies build new factories in countries overseas, such as in the United States, United Kingdom, and Asian countries.

In these lessons and activities, students engage in reading and using various kinds of distribution and land use maps. They also learn the concepts of longitude and latitude using globes (Fig. 2.3).

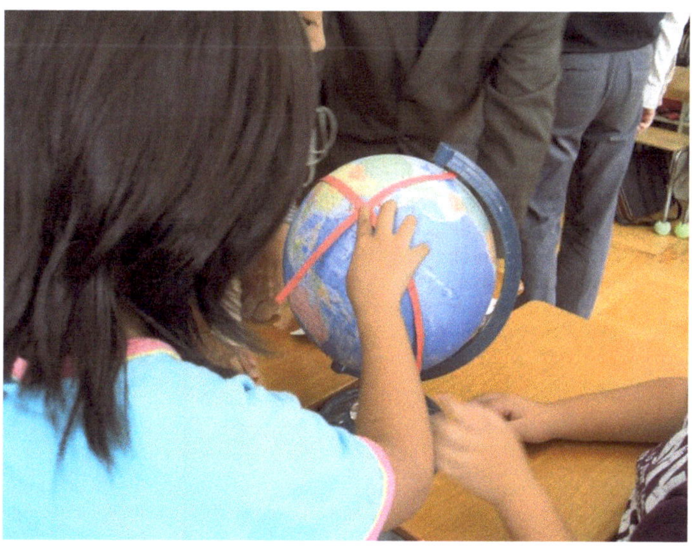

Fig. 2.3 Globe activities

2.2.5 The Sixth Grade (11–12 Years Old)

History forms a major part of the Social Studies curriculum in the sixth grade. In geography, the role of Japan in the world and the ties between Japan and other foreign countries are emphasized. Students learn about ways of life in countries that have economic or historical relations with Japan, such as the United States, China and Korea.

2.3 Perspectives on Geography Education

The subject of Social Studies in Japan is rooted in curriculum that begins at large and extends to small scale. This poses the problem that educators cannot begin to teach world geography until the sixth grade. Even though students may have a lively interest in learning about the cultures of foreign countries, the current curriculum provides little opportunity for this.

Although geography is not considered an independent subject, but a branch of Social Studies in Japan, human geography plays a more important role in the curriculum than physical geography. Therefore, elementary school students rarely have the opportunity to learn about physical geographical phenomena and the natural environment, such as landform, climate, vegetation and others.

Reform efforts at the elementary school level have produced new trends in geography education. Educators can contribute to the development of geography curricula that are grounded in contemporary approaches by continuing to test new teaching strategies and developing geographical skills and knowledge.

References

Nakayama M, Kimoto T, Komine I, Nishizawa T (1990) Chiri kyoiku no kaizen wo kangaeru: 1989 nendo shuki gakujutu takai simpojiumu I (The improvement of geographical education: symposium presentation held at the academic meeting of the association of Japanese geographers, Autumn 1989). Chirigaku hyoron (Geogr Rev Jpn) 63(Ser A)(3):166–182

Yoshida K (2008) Kodomo no asabi kodo to chikaku kankyo no hattatsu purosesu (Children's play behavior and development process of environmental perception). Chirigaku hyoron (Geogr Rev Jpn) 81(8):671–688

Chapter 3
Current Status and Issues in Junior High School Geography Education

Kaoru Hirasawa

Abstract The National Curriculum Standards for junior high schools in Japan were made public in 2008, and fully implemented in April, 2012. The key areas that were revised include the strengthening of regional geography, where students learn about the geographical characteristics of world regions and of Japan. This is combined with fostering student recognition of Japan's geographic regions and its national land, an area that has been emphasized in the geography component of junior high school Social Studies. In regards to the nation's land, the revisions call for dynamic learning in order to avoid an over-emphasis of learning facts in regional geography education. Investigative learning is another area emphasized in the new curriculum, where a survey of the local area conducted at the end of the geography field incorporates a social participation element. In order to respond to the new learning activities that come along with these revisions, the active promotion of well-planned training for educational administrative bodies, schools, and for Social Studies teachers will be needed.

Keywords Geographic skills • Social participation • The National Curriculum Standards • World recognition

3.1 Introduction

The National Curriculum Standards for junior high schools in Japan were announced in 2008 and fully implemented in April 2012, after a 2-year transition period that began in 2010. The new National Curriculum Standards retain the basic principles embraced by that of the 1998 version (Ministry of Education, Science and Culture 1999). These included firmly instilling the basic fundamentals of geography, and in the process foster the 'Zest for life' that includes studying and thinking independently.

Changes made between the two versions are in the lesson content of the geography component within Social Studies. Until now, the usual practice at the beginning of the school year in April has been to distribute textbooks to first year

K. Hirasawa (✉)
Shiraoka City Municipal Shinozu Junior High School,
2617 Shinozu, Shiraoka, Saitama 349-0204, Japan
e-mail: k-hirasawa@y8.dion.ne.jp

Y. Ida et al. (eds.), *Geography Education in Japan*, International Perspectives in Geography 3, DOI 10.1007/978-4-431-54953-6_3

students, including one each for the geography and history fields as well as an atlas, while the civics textbook is given to third year students. With the new curriculum, a new geography textbook is now distributed to second year students.[1] This new measure ensures that the newly revised geography content will be covered, replacing old textbooks that were distributed in earlier grades which would not have been sufficient for second year students.

The writing of the 2008 National Curriculum Standards was influenced by a report authored by the Central Council for Education (2008). Among the report's 'specific items for revision' in the Social Studies area, those related to the content of the geography field can be summarized in the following four points:

1. To deepen students' geographical recognition of the world, units will be added to introduce the relationship between people in various parts of the world and the environment, as well as about the diversity of the world's various regions.
2. To deepen students' further understanding of Japan's national land, content will be improved to highlight geographical characteristics that relate characteristic of regional phenomena to other phenomena across various regions.
3. To support students' geographical skills, such as reading and creating maps, these will be further emphasized throughout all of the content.
4. To foster students' competence and attitudes in solving various issues and developing the region, they are introduced to a survey of the local area.

The new school geography curriculum was based directly on these 'specific items for revision.' The following six points can be cited as key to the 2008 National Curriculum Standards (MEXT 2008):

A. Re-examination of the geography field's goals
B. Re-examination of the composition of learning content
C. Emphasis on geographical recognition of the world
D. Strengthening recognition of the nation's land through learning in dynamic regional geography
E. A greater emphasis on fostering geographical skills
F. Study on the local area that incorporates the viewpoint of social participation

The schools accept that every 10 years a revised National Curriculum Standards will be implemented. This recent introduction of standards makes big changes that affect the Social Studies and Geography courses much more than the History and Civics curriculum. Nevertheless, education groups and teachers within the Social Studies area are enthusiastically adjusting and making use of the accumulated teaching materials, developing instruction resources that align with the new content.

Educators are responding with different solutions during the transition period until the complete implementation of the 2008 National Curriculum Standards. The classes that will be taught with the fully integrated curriculum are just beginning in 2014. Keeping this point in mind, while describing the important points in the

[1] In 1969, free distribution in the elementary and junior high schools for the entire year was completed and continues up to the present.

revisions of the 2008 National Curriculum Standards, the author describes how lessons are actually being put into practice in the junior high schools in the geography field of Social Studies.

3.2 Current Status of Junior High School Geography Education

3.2.1 The Objectives and Content Structure of the Geography Curriculum

There are four objectives in the geography field:

1. About geographical recognition,
2. About grasping geographical phenomenon and regional characteristics/issues,
3. About thinking and understanding geographically, and
4. About geographical skills and attitudes.

The first objective indicates the basic goal of the geography curriculum, which under the current revisions is to cultivate students with a geographical recognition of the world's various regions (MEXT 2008).

Since the 1977 National Curriculum Standards (Ministry of Education, Science and Culture 1978), the geography instruction in junior high school has two goals: (1) to cultivate a foundation for geographical ways of looking and thinking, and (2) to foster a recognition of Japan's national land based on a broad perspective. In the 2008 version of the National Curriculum Standards, to these, 'a geographical recognition of the world's various regions' was newly added. Up until now, within the mandate of 'geographical recognition,' the 'knowledge content' of geography learning has been, in principle, to emphasize 'local area recognition' in elementary school, 'national land recognition' in junior high school, and 'world recognition' in high school. However, under the current revisions, students will learn both 'national land recognition' and 'world recognition' in junior high school geography.

Although the second objective was related to viewpoints and methods for investigating geographical characteristics in the 1998 version of the National Curriculum Standards (Ministry of Education, Science and Culture 1999), this part in the revised curriculum has been changed to emphasize geographical characteristics and regional problems. The reasoning is that the current revisions have taken into account learning on the world's geographical characteristics and Japan's regions, highlighting regional problems in the survey of the local area. This point can be seen as a lessening of the relative emphasis on 'strengthening lessons on methods of study' that were meant to instill viewpoints and methods for grasping geographical characteristics, one of the basic principles of the revisions in the 1998 National Curriculum Standards.

The third objective is indicating ways of thinking about the peculiarities and nature of geographical characteristics, followed by objective four, which is indicating the

Table 3.1 Content as indicated under the 1998 National Curriculum Standards

(1) Regional composition of the world and Japan
A. Regional composition of the world
(a) Locational relationships on the earth and distribution of land and water bodies
(b) Make-up of countries and regional divisions
B. Regional composition of Japan
(a) Japan's location and territory
(b) Make-up of prefectures and regional divisions
(2) Surveys appropriate to a region's scale
A. Local area
B. Prefectures in Japan
C. Countries of the world
(3) Japan as compared to the world
A. Japan examined from many different aspects
(a) Japan's geographical characteristics seen from the natural environment
(b) Japan's geographical characteristics seen from population
(c) Japan's geographical characteristics seen from resources and industries
(d) Japan's geographical characteristics seen from livelihood and culture
(e) Japan's geographical characteristics seen from linkages between regions
B. Japan as seen from the interrelationship of all these characteristics

Ministry of Education, Science and Culture (1999)

desirable skills and attitudes that should be instilled in students through geography learning. The intent of these two goals has carried forward to the revised curriculum and their descriptions have not been changed.

By accepting these changes in the geography field's objectives, the three content heading areas in the 1998 National Curriculum Standards (Table 3.1) [(1) regional composition of the world and Japan, (2) Surveys appropriate to a region's scale, and (3) Japan as compared to the world], have been reworked under the 2008 National Curriculum Standards into two major headings, (1) The various regions of the world, and (2) Japan's many different regions (Table 3.2).

Under the major heading (1), the topics are to form an axis for understanding the world's regions, world's regional make-up, diversity of different people's livelihood in various parts of the world, characteristics of the various regions of the world, and finally, descriptive geographical knowledge, perspective, and methods obtained through understanding world regions. The latter is learned through a process of exploration, and surveying the world's many various regions.

The major heading (2) is composed of the topics in the major heading (1), as listed above, as a strategy to develop and to deepen the lesson content and learning activities in progressive steps, based on the model of 'learning–utilizing–exploring' (Central Council for Education 2008). Through this pedagogy, students learn the fundamental vocabulary, knowledge, concepts and skills related to societal phenomena, and they are able to apply them to explore issues.

Along with the re-examination of the goals and content make-up of the geography field, changes were also seen in the number of classroom hours. Over the course of

Table 3.2 Content indicated by the 2008 National Curriculum Standards

(1) The various regions of the world
A. The world's regional make-up
B. People's livelihood and the environment in various parts of the world
C. The world's regions
(a) Asia
(b) Europe
(c) Africa
(d) North America
(e) South America
(f) Oceania
D. Surveys of the world's many different regions
(2) Japan's many different regions
A. Japan's regional make-up
B. Japan's geographical characteristics compared to the world
(a) Natural environment
(b) Population
(c) Resources and energy and industry
(d) Linkages between regions
C. Focus on Japan's regions:
(a) Consideration centered on the natural environment
(b) Consideration centered on historical background
(c) Consideration centered on industries
(d) Consideration centered on environmental problems and environmental protection
(e) Consideration centered on population and urban and rural settlement
(f) Consideration centered on livelihood and culture
(g) Consideration centered on linkages with other regions
D. Survey of the local area

MEXT (2008)

3 years, junior high school students have 350 classroom hours in Social Studies. The geography field's share of these under the 1998 National Curriculum Standards was 105 h, but under the 2008 version of the National Curriculum Standards this was increased by 15–120 h. However, the basic structure of Social Studies as a discipline has not changed. Geography and History remain compulsory for first and second year students, with an additional Geography course required in the third year, and students must study civics after their first History course.

3.2.2 *Study About the World's Various Regions*

In the School Education Law that was partially revised in 2007, among the goals of compulsory education, it is explicitly stated that, "Through an understanding of foreign cultures, to cultivate an attitude that respects other nations and contributes to the peace and development of international society". It is especially critical at the present time, with the spread of globalization, that it is extremely important to instill in students the basic and fundamental knowledge related to the world's regional diversity and to foster a geographical recognition of the whole world. Given this situation, the units focused on geographical characteristics of the world's regions have been restored in the 2008 National Curriculum Standards.

Major heading number (1) which deals with a geographical understanding of the world is divided into four sub-headings, of which three consist of newly established content. The first of these is '(B) People's livelihood and the environment in various parts of the world'. This can be seen as responding to the 'emphasis on learning about many different traditions, cultures, and religions' that was spelled out as one of the basic principles in the revision of junior high school Social Studies, by more or less restoring the content of 'people's livelihood and the environment' from the 1989 version of the National Curriculum Standards (Ministry of Education, Science and Culture 1989).

The second is '(C) The world's regions'. In this heading, geographical phenomena provides a context through which one can solidly grasp the pattern of the livelihood of people living in the six major regions of Asia, Europe, Africa, North America, South America, and Oceania. The learning is positioned as regional geography education that will support students' understanding of the geographical characteristics of these regions. In the interpretation of the 2008 National Curriculum Standards, 'topics' are themes to be decided by the teacher while considering other factors, such as the geographical phenomena being dealt with, the content previously covered, the ease or difficulty of the topic, the student's life experience, the projected learning activities, and the classroom time involved. Teachers will need to consider whether there is sufficient time to cover the proposed topics as well as teach an overview of each region's geographical characteristics from many different aspects. For that reason, one should focus on only one or two topics for each major region. Also, when setting out the topics, it is necessary to decide whether they lend themselves to drawing comparisons or relationships with Japan, and whether the pattern of human life in the world's regions are seen from the many aspects. Topical study is excellent in that one can pick and choose from among the immensity of geographical phenomena. However, the main emphasis should be on making geographical themes clear, so if a mistake is made in the choice of topics then the regional nature will be unclear (Kobayashi 2009).

The third sub-heading with new content is '(D) Surveys of the world's many different regions'. Although only partially, the content of 'instilling the perspectives and methods for grasping geographical characteristics,' which is emphasized in both the 1989 and the 2008 National Curriculum Standards.

As far as the topics for the six major regions are concerned, many examples from the National Curriculum Standards' interpretation have been incorporated in the textbooks, so lesson plans are created in alignment with these. However, the order in which they are taught differs across teachers. For example, at the 2010 National Junior High School Social Studies Education Research Association conference in Sendai-Miyagi, a lesson plan proposed to begin the course with North America first.

The rational takes into consideration the students' developmental level and asserts that learning about Asia, which is so broad, complex, and diverse as the first of the 'regions of the world' would be difficult for them. In contrast, North America was seen as a coherent region and an easy one to compare industrial models. The lesson proposed to then move to Asia, Europe, Oceania, and then to Africa. The author planned Africa to be taught last because it was reasoned that it would be difficult for students to learn about African monoculture so early on in the first year of junior high school.

Then again, at the 2011 National Junior High School Social Studies Education Research Association's Tokyo conference, in focusing on whether or not it would be easy for students to construct an image of the world, a lesson plan was proposed that began with the Oceania region (Tokyo Metropolitan Junior High School Social Studies Education Research Association 2011). The interest level and concern for Australia, which is a core region of Oceania, would be high among the students making it easy to show the relationship between people's livelihood and the natural environment. For this reason, the author thought it appropriate to deal with Oceania at the beginning of the 'world's various regions,' and focus on immigration as a topic. The lesson moves on to Europe, North America, South America, and then to Africa. After learning about Africa, students would study the Asian region and the range of different climates, ethnicities, cultures under the topic of 'population problems.' Having outlined the geographical characteristics of the world's various regions, the lesson focuses on Japan and its local regions.

3.2.3 Study on the Various Local Regions of Japan

In geography, the study of Japanese regions has tended to fall into an encyclopedic treatment of each heading, and moreover there has been a tendency for students to memorize facts about the regions being studied. In the 1998 National Curriculum Standards, it aimed to strengthen the ways of studying such as by investigating topics in classes in order to move away from memorization, rather towards nurturing students' abilities to respond to the changes in society. In contrast, the 2008 National Curriculum Standards emphasize the acquisition of basic and fundamental knowledge and skills as a means of cultivating a base in geographical ways of thinking. This will be achieved through students pursuing and explaining the relationships between phenomena. From this perspective, study on the various local regions have been revived. What is needed, after dividing Japan into regions, is the development of dynamic regional geography learning that help students grasp the geographical characteristics by focusing on key characteristic phenomena in each of the regions and relating them to other phenomena.

Under the major heading that deals with the geographical recognition of Japan, (2) 'Japan's many different regions,' a new sub-heading has been established, (C) 'Japan's regions.' The content of this is to divide Japan into a number of regions, and to help students grasp their geographical characteristics based on seven different means of considerations or avenues of approach to each of the regions: (a) the natural environment, (b) the historical background, (c) industries, (d) environmental problems and environmental protection, (e) population and urban and rural settlement, (f) livelihood and culture, and (g) linkages with other regions.

Each of the seven consideration or approaches should be chosen for each of the regions studies, which means that at a minimum Japan should be divided into seven regions. Moreover, taking into account the instructional perspective and geographic location of the school, educators could conceivably divide the country into more than seven regions, but in whichever case, teachers are directed to be able to deal with all seven of the approaches (a) through (g) in the course of covering all the regions of Japan.

In studying Japan's local regions, it is necessary to first think about the regional divisions and the order in which the regions should be taught. General textbooks and other materials often begin with the Kyushu region (including Okinawa) and proceed northward, in order. However, the National Junior High School Social Studies Education Research Association (2010), at their 2010 Sendai-Miyagi conference, proposed a lesson plan in which the order of instruction begins with the Tohoku region, and proceeds through the regions of Kyushu, followed by Chugoku, Shikoku, Kinki region, Chubu region, Hokkaido, and Kanto, before finishing with a survey of the local area. The order in which regions are taught follows past tradition where the study of the local area came before the study of the other regions of Japan. From that period on, with the perspective of creating a linkage between regions, the instruction of the various regions commenced with the Tohoku region. The study of Japan's regions ends with Kanto since it is where Tokyo, the destination for most school study-trip excursions, is located. This order was proposed giving consideration to that historical background and regional circumstances.

Next, in order to clarify the characteristics of the individual regions, it is important which geographical phenomena are chosen as the core topic. At present, there are four companies that publish geography textbooks.[2] Table 3.3 shows the combinations of each of the seven approaches along with the regions in which they are associated with in the various textbooks. The Tokyo Metropolitan Junior High School Social Studies Education Research Association (2011), proposed the combination that matches example 1 in Table 3.3.The purpose is to heighten students' interest and enthusiasm while guiding their grasp of geographical characteristics. To begin with, the Kyushu region would be studied through an approach centered on (a) the natural environment, which is the basis of the geographical way of looking and thinking. Finally, the Kanto region, where one third of Japan's population is concentrated is dealt with under (e) population and urban and rural settlement. The rational for putting the Kanto region last among the 'various regions of Japan' is that by studying about a region with a large settled area last, the geographical characteristics and issues that were highlighted there could be organically tied to

[2]The four publishers of junior high school Social Studies (with geography content) textbook passed the 2011 textbook approval process are: Tokyo Book Publishing, Inc.; Educational Publishing, Inc.; Teikoku Shoin, Inc.; and Nihon Bunkyo Publishing, Inc.

Table 3.3 Combinations of geographical approaches and regions to be addressed in textbooks

Geographical approach perspective	Example 1	Example 2
(a) Natural environment	Kyushu	Hokkaido
(b) Historical background	Hokkaido	Kinki
(c) Industries	Chubu	Chubu
(d) Environmental problems and environmental protection	Kinki	Kyushu
(e) Population and urban and rural settlement	Kanto	Chugoku & Shikoku
(f) Livelihood and culture	Tohoku	Tohoku
(g) Linkages to other regions	Chugoku & Shikoku	Kanto

Based on the textbooks of four publishers (see footnote 2)

the 'survey of the local area' to create a close continuity between 'Japan's various local regions' and the 'survey of the local area.' Also, having piqued students' interest and increased understanding in the learning content, it was thought that it would be possible to carry out a 'survey of the local area' and that this would be effective as well in 'nurturing a consciousness of social participation among the students,' simultaneously reaching the aims of 'Education for Sustainable Development' (ESD) in creating a sustainable society.

Ushigome (2009) proposed three methods to teach the core phenomena of geography topics: (1) to focus on all core phenomena, 2(2) to focus on select core phenomena, and (3) to use "window" method by studying phenomena of human and physical geography topics across different regions.

3.2.4 *Study of the Local Area*

The Basic Act on Education and the School Education Law, partially revised in 2007, clearly states that 'an attitude will be cultivated by participating as an actor in the formation of society and contributing to its development, based on public spiritedness.' The nurturing of a social participation attitude also means to cultivate a foundation of civic qualities, a goal of Social Studies. Learning geography also deepens an understanding and interest to participate in the development of the local area where students live. This objective is under the content subheading (2)–(D) 'survey of the local area' in the 2008 version of the National Curriculum Standards. It emphasizes investigative learning that incorporates social participation on local regional issues in students' communities giving them opportunities to participate in local community formation and to cultivate an attitude of students' desire to work for their region's development. 'Local regional issues' are defined as geographic phenomena with a location or an areal extent. It is necessary to distinguish these from local regional issues such as regional fiscal budgets, etc. that are related to content taught in civics classes.

Classes, not limited to geography, emphasize the incorporation of surveys and observation, study trips, and experience. However, according to the survey results[3] of Research on Specific Issues (Social Studies), only 20.8 % of schools confirmed that field work and questionnaire surveys were included in the topic 'survey of the local area' (National Institute for Educational Policy Research – Curriculum Research Center 2008). The reasons that local area surveys are not carried out very much, as seen from the standpoint of the schools, is that teachers are overburdened with work in addition to their instructional duties. Moreover, in the high school admissions exams, it is difficult in reality to pose questions related to the 'survey of the local area,' so questions related to geography are usually limited to those testing the ability to read topographic maps.[4] Because of this testing format, many teachers focus on teaching map reading in the classroom, another reason that local area surveys are not often carried out.

As a way of promoting the 'survey of the local area,' Matsuoka (2010) leads a program called 'One-Point Geographical Excursions.' It is a geography education excursion focusing on select phenomena in the vicinity of the school, equivalent to a one credit hour. Through this implementation, the exercise leads to an increase of students' observation skills and fosters their interest to be aware of local phenomena. The goal is that students will independently explore similar phenomena outside of school.

From the perspective of strengthening language-related activities, Harigaya et al. (2008) conducted a survey of the local area using maps and in the process devised activities where students could collect and communicate their findings from the exercise. In this case, when presenting the results of the survey, they create a 'guide map to introduce the area' where students make an outline map, with catch phrases that are explained, of the administrative divisions at the scale of city, town, and village. In order to make active use of geographical and communication skills, a series of presentations titled 'Rediscovery of the local area through presentations' provide a summary of the field survey.

The Saitama Prefecture Social Studies Education Research Association (2009) have proposed putting into practice activities that incorporate social participation. One example asks students to imagine their local community 30 years in the future and what they want it to be. Students present the results of their imagined local area survey, based on projections of the issues, illustrated on a map. Through this project, students learn to take notice of the positive features and issues of the local area, and it is an attempt to heighten their awareness through participation in the society. As an example to bring about awareness of societal participation, the annual teaching plan of the Tokyo Metropolitan Junior High School Social Studies Education Research Association (2011) are trying to

[3] Made public in June, 2010; survey conducted on January 24, 2009; 9,394 third-year junior high school students at 301 schools participated.

[4] According to the *Analysis of National High School Entrance Exam Questions for 2013 Exam Takers—Social Studies*, published by Obunsha Press, 27 out of 47 prefectures had questions related to topographic maps on the 2012 high school entrance exams.

understand the thinking of forebears who created a better community and heightened interest in the local area through studying the boundary adjustments that were made in the past.

3.2.5 The Nurturing of Geographical Skills

In the geography field, the importance of utilizing large scale maps and atlases has constantly been pointed out. Simply, maps as tools, map reading, and map making are indispensable skills for understanding geographical phenomena and for advancing the fundamental geographic ways of looking at and thinking about the world. By using maps effectively to explain phenomena, in the meanwhile adding one's own interpretations and making arguments, and exchanging ideas, are all elements to fostering skills (e.g., abilities to think, make judgments, and to express oneself) unique to map learning. From this standpoint, along with Social Studies courses in elementary school and Geography and History in high school, to classes in the geography field of junior high school Social Studies, much emphasis is put on nurturing geographical skills that are focused on the utilization of maps.

Geographical skills, including the geographic way of looking and thinking, are not something that can be instilled through one class or a single experience, but rather developed through the process of repeated learning in which the level of familiarity gradually increases. Towards that end, as teaching materials intended to foster geographical skills, the Saitama Social Studies Education Research Association (2009) proposed the creation of study cards. First, they created a list of the basic knowledge, concepts, and skills related to the geographical expectations outlined. Then, as teaching materials that would foster the geographical skills on this list, they created study cards in color that explained, for example, the reading and interpretation of rank-division maps, or the methods for creating a population graph.

Harigaya et al. (2008) attempted to systematically arrange learning about maps and topographic maps in elementary, junior high and high schools by focusing on three points: 'learning content and aims,' 'how to read maps and express things in maps,' and 'suggestions for the use of maps and topographic maps.' However, while the importance of similar mapping skills have been pointed out (e.g., Ida (2008)), the curriculum for learning these skills at four different levels, from elementary through high school, has yet to be implemented. The learning of skills in incremental stages is something that should be implemented in the future.

Given the advancement of globalization, increasing geographical recognition, and the enhancement of geographical skills, it is now more important to cultivate the basic resources and skills so students can appropriately make use of available information and information technology. From this standpoint, it is desirable to actively devise ways of utilizing computers and information and communications networks. The Fujisawa City Education and Culture Center (2007) leads training with mapping information acquired from geographic information systems (GIS) for the creation of maps and graphs. They are assisting in the implementation of classes at each of the elementary and junior high schools in the city.

3.3 Future Issues

This chapter described the important points of the current revisions in geography as well as the current status of geography education. A number of issues remain. To begin with, due to the newly added content in the current revisions, the increase in learning content is greater than the increase in the number of hours that are allotted for. In order to prevent a shortage of hours from creating the impression that 'geography is a memorization subject,' it is necessary to further advance the development of teaching materials and the improvement of teaching methods and strategies. For this reason, it is particularly important to promote active learning and to strengthen learning that deals with relevant issues to student interest in order to further cultivate their abilities in problem solving. Promoting such learning enhances their thinking, judgment, and expressive abilities which are necessary to confront issues of the twenty-first century and beyond.

A pressing taskis creating ways to promote teacher training. Based on teachers' geographic learning from the courses they have taken, those in their 30s and younger did not study about world regions when they were in junior high school. To mediate teacher knowledge, models of professional training exist. For example, Saitama Prefecture provides newly hired teachers 30 h of training but among those, no more than 3 h are dedicated to field work (Saitama Prefecture General Education Center 2012). Moreover, the total number of teachers in Social Studies who majored in geography is small, relative to history, politics or economy. One can easily see how these factors contribute to low implementation of map use, outdoor observation, and local area surveys. The current issues are compounded by limited teacher initiatives such as asking other teachers at the workplace for instructional materials or voluntarily participating in the study training courses offered by research groups or academic associations.

Given the demands on teachers, it is necessary for the educational administrative bodies to actively promote the development of teaching materials and systematic training on the knowledge and skills important to geography field instruction (e.g., map reading and local area surveys) as well as to further increase the number of teachers who enjoy teaching geography. At the school level, it is important to build a collegial environment that facilitates the sharing of the latest research through courses, to include external support and resources, to create and introduce new teaching materials. In the future, teachers should actively participate in in-service training to learn new geographic skills and teaching content as well as develop confidence to reconstruct new and existing materials that conform to the National Curriculum Standards.

References

Ministry of Education, Science and Culture (1999) Chugakko gakushu sido yoryo (Heisei 10 nen 12 gatsu) kaisetsu shakaihen (Junior high school course of study (December, 1998) – explanation– social studies). Osaka shoseki, Osaka

Central Council for Education (2008) Shogakko, chugakko, kotogakko oyobi tokubetsu shien gakko no gakushu shido yosho no kaizen ni tsuite (toshin) (On improvements to the course of study for elementary, junior and senior high schools and special assistance schools (Report). http://www.mext.go.jp/a_menu/shotou/new-cs/news/20080117.pdf. Accessed 26 Dec 2013

MEXT (2008) Chugakko gakushu shido yoryo kaisetsu shakai hen (Junior high school national curriculum standards– explanation– social studies). Nihon Bunkyo Publishing, Tokyo

Ministry of Education, Science and Culture (1978) Chugakko shidosho shakaihen (Junior high school teaching manual– social studies). Osaka shoseki, Osaka

Ministry of Education, Science and Culture (1989) Chugakko shidosho shakaihen (Junior high school teaching manual– social studies). Osaka shoseki, Osaka

Kobayashi M (2009) Chiri gakushu ni okeru shudai gakushu to doutai chishi no saikentou: shudai gakushu no saiteigi to tomoni (A study on the thematic geography and dynamic regional geography in geography teaching– together with a redefinition of thematic geography). Chiri kyoiku kenkyu (J Geogr Edu) 3:9–17

Tokyo Metropolitan Junior High School Social Studies Education Research Association (2011) Atarashii gakushu shido yoryo ni motozuku bunya betsu shido keikaku (Lesson plans by section based on the new course of study). Tokyo Metropolitan Junior High School Social Studies Education Research Association, Tokyo

National Junior High School Social Studies Education Research Association (2010) Dai 43 kai zenkoku chugakko shakaika kyoiku kenkyu taikai: Miyagi, Sendai taikai, kenkyu kiyo (Research reports of the 43rd National Junior High School Social Studies Education Research Conference in Sendai-Miyagi). Zenkoku chugakko shakaika kyoiku kenkyukai (National Junior High School Social Studies Education Research Association), Tokyo

Ushigome H (2009) Chugakko ni okeru nihon no shochiiki gakushu no hoho (Methods for Japan regional study in junior high school). Chiri kyoiku kenkyu (J Geogr Edu) 3:20–22

National Institute for Educational Policy Research- Curriculum Research Center (2008) Tokutei no kadai ni kansuru chosa (shakai) chosakekka (shogakko, chugakko) (Research on specific issues (social studies)–survey results (elementary and junior high schools). National Institute for Educational Policy Research, Tokyo

Matsuoka M (2010) Chiri kyoiku ni okeru junken gakushu ron no kochiku to wan pointo junken no teisho (On the theory of geographical excursion studies and one- point geographical excursions). Chiri kyoiku kenkyu (J Geogr Edu) 7:1–7

Harigaya S, Mita H, Ono T (2008) Sho chu ko no chiri gakushu no keitousei to hattensei: chizu, chikeizu gakushu wo tooshite (Systems and possibilities of geography study in primary schools and junior and senior high schools–through the study of maps and topographic maps). Chiri kyoiku kenkyu (J Geogr Edu) 1:19–26

Saitama Prefecture Social Studies Education Research Association (2009) Heisei 21 nendo dai27 kai kanto burokku chugakko shakai ka kyoiku kenkyu saitama taikai kenkyu kiyo (Research reports of the Saitama Conference, 2009 27th annual Kanto block junior high school social studies education research conference). Saitama Prefecture Social Studies Education Research Association

Ida Y (2008) Chiri gino ni yoru ikkan karikyuramu (A consistent curriculum through geographic skills). In: Yamaguchi S et al (eds) Chiri kyoiku karikyuramu no sozo (The creation of a geography education curriculum). Kokon shoin, Tokyo, pp 109–115

Fujisawa City Education and Culture Center (2007) Atarashii chiiki gakushu: shiminsei no ikusei, kyoiku GIS wo riyo shite (New local area study– fostering citizenship– using an Educational GIS). Fujisawa City Education and Culture Center, Fujisawa

Saitama Prefecture General Education Center (2012) Heisei 24 nendo sho chugakko shoninsha kenshu no tebiki (2012 training manual for newly hired teachers in elementary and junior high schools). Saitama Prefecture General Education Center, Saitama

Chapter 4
Current Status and Issues in Senior High School Geography Education

Toshio Asakawa

Abstract Patterns of enrollment and learning content in senior high school geography education have changed since the new postwar system was launched. Not all senior high school students are necessarily currently studying geography. In the new elective subject entitled Geography A, they study geographic characteristics, problems of the contemporary world, and geographic issues in the living environment. In Geography B, students study a variety of maps and learn geographic skills, systematic geographic topics, and descriptive geography in the contemporary world.

Keywords Geography education • Learning content • Pattern of enrollment • Senior high school

4.1 Changes in the Pattern of Enrollment and Learning Content

Unlike at the elementary and junior high school levels, not all senior high school students are necessarily studying geography. The pattern of geography-related subject enrollment within the senior high school curriculum has changed three times since the post-war education system was launched in April, 1948. During the first 15-year period, "Human Geography" was established as a geography-related subject; it was an elective subject together with History- and Civics-related courses. For a 20-year period beginning in 1963, Human Geography was divided into two subjects, entitled Geography A and Geography B, and it was compulsory for students to enroll in one or the other. In the early implementation of these two courses, they were different insofar as the number of instruction hours. However, from 1973 onward, as described below, the learning content also differed. From 1982 to the present, having unified systematic geography and descriptive geography into a single subject simply called "Geography," the existing courses Geography A and

T. Asakawa (✉)
Saitama Prefectural Urawa Daiichi Girls' Senior High School,
3-8-45 Kishicho, Urawa, Saitama 330-0064, Japan
e-mail: asakawa-sanputei@ac.wakwak.com

Y. Ida et al. (eds.), *Geography Education in Japan*, International Perspectives in Geography 3, DOI 10.1007/978-4-431-54953-6_4

Table 4.1 Changes in enrollment and learning content patterns

Year	Pattern of enrollment	Subjects	Learning content
1948	Elective subject	Human geography	Systematic geography
1963	Elective subject	Geography A	Systematic geography (four units)
		Geography B	Systematic geography (three units)
1973	Compulsory subject	Geography A	Systematic geography
		Geography B	Regional geography
1982	Elective subject	Geography	Systematic geography
1992	Elective subject	Geography A	Topical methods
		Geography B	Systematic geography
2003	Elective subject	Geography A	Methodological abilities
		Geography B	Case studies

and Geography B, whose organizing principles vary regarding learning methods and learning content, have become elective subjects. In particular, since the category of Social Studies, which had continued since the beginning of the new senior high school system, was divided in 1992 into Geography and History and the Civics categories. Within the Geography and History category, only World History-related subjects have been mandatory; Geography-related subjects have become optional, students choosing between Geography and Japanese-history related subjects.

More or less in parallel with alterations in the pattern of enrollment, course learning content has also changed. A brief outline of these changes is described in Table 4.1, following up on the work of Nishiwaki (2009).

Learning content of the first geography-related subject introduced under the new post-war senior high school system, Human Geography, focused on answering the question, "How do human beings live on the earth's surface and build cities and villages?" (Ministry of Education, Science and Culture 1951). Learning systematic geography was emphasized, which revolved around productive activities. Efforts were made to clarify the systematic geography learning content by arranging it into topics such as "farming and herding" and "settlements". Content related to maps and outdoor field surveys was added to this Human Geography subject, and geographic skills were also emphasized. In the next stage, the learning content of Geography A and Geography B also focused on Human Geography, protection of the natural environment, and Japan's land area. There was little difference between teaching content of the two subjects; they only differed in the number of instruction hours, with Geography A being three credit units and Geography B four credit units. In the 1973 version of the educational curriculum, however, the teaching content in Geography A and Geography B classes differed. The content of Geography A focused on the existing systematic geography, that of Geography B centered around world regional descriptive geography, studying world regions through such topics as "characteristics of the natural environment," "inhabitants and population," and "the current state and trends in industry and the economy" (Ministry of Education, Science and Culture 1970).

The new subject entitled Geography, launched in 1982, was established in response to the shift in enrollment pattern caused by changing geography from a compulsory to elective subject. The learning content, which united systematic geography and world regional descriptive geography while maintaining a largely systematic geography structure, also incorporated local regional methods and topical learning methodology. From 1992 onward, along with establishment of the Geography and History discipline category, new versions of Geography A and Geography B were established. Not only did these two subjects differ according to hours of instruction, with A having two credit units and B four, the learning content and the organizing principles varied substantially. In Geography A, learning based on topical methods was emphasized, and the content was about understanding other cultures and considering local issues in a regional way. Conversely, in Geography B, both methods and content were carried over from the existing systematic geography content. Despite content differences, both subjects emphasized instruction based on case studies and concrete examples, rather than current factual knowledge that would soon become obsolete in a rapidly changing society. Emphasis was placed on student development of methodological abilities, pushed forward even more strongly by Geography A and Geography B from 2003 onward. The new Geography A, although emphasizing the development of methodological abilities, essentially repackaged the learning content of the old Geography A. In contrast, through use of case studies to acquire knowledge of systematic geography and regional descriptive geography, the content of Geography B took on the core of systematic geography and descriptive geography, which included issues of the contemporary world. While adopting the educational curriculum that has been gradually implemented each succeeding school year beginning in 2003, students study systematic geography and regional descriptive geography until these subjects are replaced by the new Geography A and Geography B described below.

4.2 Enrollment Trends of Recent Years and Expected Changes

Nearly 30 years have passed since geography-related subjects were compulsory. During that time, the National Curriculum Standards have been revised four times. The education curriculum has been restructured, and each time there has been a movement by many in geography education asking for a reexamination of subjects in which students enroll. The new education curriculum, put into effect in April, 2013, is based on the National Curriculum Standards that were revised in March, 2009. Despite these efforts, there have been no changes in the enrollment status of Geography and History courses. For the near future, we can expect the situation to continue as is, with World History-related subjects as required classes and Geography-related and Japanese History-related subjects as elective.

Geography-related subjects became optional in 1982. From the number of geography textbooks supplied and number of students taking geography exams at the National

Center preliminary university entrance examinations, geography-related subjects occupy a uniform position in the education curriculum, and it has played a consistent role in senior high school education (Asakawa 2006).

This trend has not changed in recent years. For example, the textbook supply rate (number of textbooks supplied divided by total number of students times 100) in 2012 is almost unchanged compared to 1997 (when the education curriculum initiated in 1992 was completed) or from 2006 (when the education curriculum initiated in 2003 was completed). Using the textbook supply ratio as a comparison to World History-related subjects that are required for all senior high school students and interpolating enrollment levels in Geography-related subjects in recent years, the figure comes to about one-half that of World History subjects. So, even considering that some students elect to take both Geography A and B, one can assume that roughly half of senior high school students are enrolled in any geography classes. The situation is even dire when we consider that among Geography-related subjects, the proportion of students taking Geography B (four credit units) has fallen. The number taking Geography A (two credit units), which is easier to combine into the overall education curriculum, has increased. Thus, one can conclude that the number of hours in which students are enrolled in geography subjects is declining (Table 4.2).

In thinking about the future of senior high school education, we foresee two changes in enrollment trends. One scenario is a potential decline in the number of students taking geography, compared with an increase of students taking Japanese history-related subjects. In turn, this will reduce the number of geography teachers needed.

As Niihori (2006) points out, we are in a period when large numbers of teachers responsible for senior high school geography education are retiring. Moreover, the number of newly hired teachers who have a specialization in geography is extremely small, and there is a serious shortage of geography subject teachers. Tables 4.3 and 4.4 show the decline in student enrolment in geography subjects and a small number of geography teachers hired, in Saitama Prefecture.

The other scenario is the potential increase in Geography-related course enrollment due to students choosing Geography and History, and Civics subjects in the National Center preliminary university entrance examinations (hereafter referred to as "National Center exams").

Table 4.2 Textbook supply ratios for subjects in Geography and History

	World History subjects		Japanese History subjects		Geography subjects	
Year	World History A (%)	World History B (%)	Japanese History A (%)	Japanese History B (%)	Geography A (%)	Geography B (%)
2012	26.7	14.5	12.6	16.3	12.6	8.1
	41.2		28.9		20.7	
2010	26.0	15.1	12.1	16.2	13.1	7.5
	41.1		28.3		20.6	
2006	–	–	–	–	12.3	8.1
	39.6		27.1		20.4	
1997	–	–	–	–	11.5	9.4
	37.6		26.2		20.9	

Table 4.3 Status of subject offerings and enrollment in Japanese History and Geography in Saitama Prefecture

	Japanese History subjects		Geography subjects	
Year	Schools offering (%)	Student enrollment (%)	Schools offering (%)	Student enrollment (%)
2006	92.2 (177 schools)	75.1	83.4 (160 schools)	52.6
2009	95.0 (171 schools)	78.6	76.6 (138 schools)	49.7

Table 4.4 Number of Geography and History teachers hired in Saitama Prefecture

	Hiring Year					Total for the period
	2006	2007	2008	2009	2010	
World History	1	0	2	3	2	8
Japanese History	1	3	0	2	8	14
Geography	0	0	0	1	2	3
Yearly Total	2	3	2	6	12	25

Excludes those hired by special assistance schools

Table 4.5 Number of students taking exams in Geography and History and Civics in the National Center exams for University Admissions and choice ratios

	2011		2012	
Subject	Exam takers	Choice ratio (%)	Exam takers	Choice ratio (%)
World History B	88,303	16.7	91,139	17.3
Japanese History B	152,970	29.0	157,372	29.9
Geography B	113,769	21.6	132,528	25.2
Contemporary Society	177,843	33.7	105,570	20.1
Ethics	58,274	11.0	35,537	6.8
Politics and Economy	88,758	16.8	57,224	10.9
Ethics and Politics and Economy	–	–	49,601	9.4
Number of Main Exam Takers	527,405		526,311	

Choice ratio = exam takers per subject/main exam takers × 100

Students could choose to take one subject exam from each of two categories, Geography and History, and Civics. From 2012 onward, however, these two subject areas have been combined into one group. This new Geography–History–Civics subject gives students ten subjects from which to select two for the exam. Because of this change, top national and public universities limit students from taking exams for "A" subjects with two credits in Geography and History, and the subjects Contemporary Society, Politics and Economy, and Ethics in the Civics category. As a result, the choice ratios for Geography and History and Civics (number of students taking each subject exam divided by the total number taking the general exam times 100) showed a large decline in the Civics subjects in the 2012 National Center exams, compared with 2011 data. By contrast, the choice ratios showed a conspicuous increase in Geography B within the Geography and History category (Table 4.5).

Creating a curriculum that responds to students' desire to take Geography B instead of Contemporary Society or Politics and Economy is being considered. This is especially true for students at senior high school level who are applying to science-related universities, but who until now have been choosing to take exams in Contemporary Society or Politics and Economy within the National Center exams.

4.3 Learning Content of the New Geography A and Geography B

High school administrators are uncertain which of the two described scenarios may occur. They are required to respond to the new learning content stipulated in the officially published revised national curriculum of March 2009.

In the 2009 national curriculum (MEXT 2009), two new geography subjects were established, the new Geography A (two credits) and Geography B (four credits). With respect to learning content, the new Geography A is composed mainly of thematic subject methods, and the new Geography B primarily of systematic geography and descriptive regional geography methods. In the basic framework of their organizational content, there are no differences between the new and existing Geography A and B. However, some new characteristics and a number of differences are evident in specific learning content (Table 4.6). One of these, common to both the new Geography A and Geography B, is the goal of advancing student learning with maps. Particularly in Geography B, learning content related to maps and geographic skills is established from the very beginning under a separate heading "All kinds of maps and geographic skills," which is positioned as the basis for subsequent content. Another characteristic shared by Geography A and B is an emphasis on learning that explores geographic issues. At the end of the new Geography A course, students investigate geographic issues in the living environment using learning accumulated to that point, in the module called "Geographical issues of the living environment and field surveys." In "The contemporary world and Japan" at the end of the new Geography B, students investigate Japan's various geographic problems by applying knowledge learned.

In addition to these, new characteristics and differences are seen in each subject. In the new Geography A, the aim is to strengthen learning content that relates to everyday life and emphasizes the utility of geography. Specifically, together with existing content related to the contemporary world, content that deals with issues of the living environment as considered from a geographic perspective was newly added under a separate major heading, "Geographical consideration of various issues in the living environment". Within this initiative, learning about maps of the immediate surroundings ("Maps that are tied together with everyday life"), about disaster prevention in response to the wave of earthquake disasters ("The natural environment and disaster prevention"), and similar topics have been added. In the new Geography B, the goal is to provide learning content that emphasizes the contemporary world, in which regional descriptive geography is strengthened. Current descriptive geography learning focuses on two or three exemplary regions, depending

Table 4.6 Learning content stipulated by the National Curriculum Standards of 2009

Content of Geography A
1. Geographical consideration of characteristics and various problems of the contemporary world
a. The contemporary world seen in globes and maps
b. Diversity of the world's living patterns and cultures
c. Geographical consideration of global problems
2. Geographical consideration of various issues in the living environment
a. Maps linked to everyday life
b. Natural environment and disaster prevention
c. Problems in the living environment and local region survey
Content of Geography B
1. All kinds of maps and geographic skills
a. Geographic information and maps
b. Map use and field survey
2. Systematic geographic consideration of the contemporary world
a. Natural environment
b. Resources and industries
c. Populations, cities, rural settlements
d. Living pattern, cultures, ethnicity, and religions
3. Descriptive geography of the contemporary world
a. Regional divisions of the contemporary world
b. Various regions of the contemporary world
c. The contemporary World and Japan

Based on MEXT (2009)

on their scale, from among the regions of the world. The new Geography B approach diverges where content has been added to treat global regions of various scales, in a standardized way ("Various regions of the contemporary world"). Further, along with more robust learning content, there is significant change in the position of descriptive regional geography in the new Geography B. That is, until now, descriptive geography learning together with systematic geography learning has been positioned as a tool or method for considering issues of the contemporary world. The new Geography B—based on maps and geographic skills, systematic geographic method learning, and descriptive geography considering regional characteristics or problems of regions—has now been positioned as a learning "goal" or a "summing up" of student learning. Accompanying this change in positioning of descriptive geography learning, the treatment of learning content related to issues of the contemporary world is also changing in Geography B.

Until the latest curriculum revisions, learning contents were categorized separately. However, they are now grouped into categories of topics studied using systematic geography methods ("Systematic geographic consideration of the contemporary world") and those studied with descriptive geographic methods ("Descriptive consideration of the contemporary world"). In the systematic geography units, students learn about phenomena from a global perspective, so as to sufficiently understand regional characteristics and issues to think independently about problem solutions.

4.4 Issues for the Future

Three issues are highlighted from the examination of the current status of senior high school geography education.

The first is the response to learning that explores various geographic problems, which are emphasized in both the new Geography A and Geography B. Although it is difficult to maintain that learning incorporates local regional surveys or that exploratory activities have been adequately carried out through the present (National Education Policy Research Institute 2007), many teachers have been confused over how to embody and implement the new learning content. Several urgent needs remain: (1) To examine outcomes and problems in cases where the new learning content has already been tackled; (2) to share examples of practice through publications and the Internet; and (3) to develop and provide relevant teaching materials created through collaboration between researchers and educators in the schools.

The second issue is the development of teaching materials that continue to demonstrate to students the usefulness of geography. Students have reported the usefulness of geography subjects in comparison with other subjects within the Geography and History categories (National Education Policy Research Institute 2007). Students become aware of geography's value by learning about relevant topics such as disaster prevention; hence, these have been added to the content in the 2009 National Curriculum Standards. Another tactic has been to foster exploratory learning, bringing together disaster prevention researchers as an example, to collaborate with teachers in developing specific teaching materials in conformity with the content. These resources are available to the larger education community and create a strong argument for the utility of geography, in the hopes of increasing student enrollment. Foreseeable obstacles include a continuing decline in the number of teachers specializing in geography, coupled with finding teachers who can teach geography subjects. The number of qualified teachers with geography skills (e.g., map use, the natural environment, and conducting surveys) needs to increase in order to halt the decline in geography classes being offered. One challenge is that good training opportunities are not provided for teachers of Geography and History, so they tend to shy away from teaching geography subjects. Thus, geography teachers are a critical element of the solution to increase geography enrollment numbers in the future.

References

Asakawa T (2006) Chirikei kamoku no rishu doko (Enrollment trends in geography subjects). Chiri 51(8):56–59

MEXT (2009) Koto gakko gakushu shido yoryo, Heisei 21 nen 3 gatsu (School curriculum guidelines for senior high schools, March 2009). Higshiyama shobo, Kyoto

Ministry of Education, Science and Culture (1951) Chugako kotogakko gakushusdo yoryo shakaika hen III (c) Jinbun chiri (shian) (The 1951 revised edition of the national curriculum for junior and senior high schools, social studies section, 'human geography' draft). Ministry of Education, Science and Culture, Tokyo

Ministry of Education, Science and Culture (1970) Koto gakko gakushu shido yoryo (National curriculum for senior high schools). Ministry of Education, Science and Culture, Tokyo

National Education Policy Research Institute (2007) Heisei 17 nendo kyoiku katei jissshi jokyo chosa (koto gakko) chosa gaiyo, shukeihyo: chiri rekishi (Survey of implementation status of the 2005 education curriculum (senior high schools) – summary of results and statistics–geography and history). National Education Policy Research Institute, Tokyo

Niihori T (2006) Chiri ga nakunaru? Chiri kyoshi ga inaku naru? (Will geography disappear? Will geography teachers disappear?). Chiri 51(6):22–27

Nishiwaki Y (2009) Gakushu shido yoryo to kyokasho no hensen (The national curriculum and changes to textbooks). In: Nakamura K, Takahashi N, Taniuchi T, Inui T (eds) Chiri kyoiku koza dai1kan: chiri kyoiku no mokuteki to yakuwari (Course in geography education, vol. 1. The goals and role of geography education). Kokon shoin, Tokyo, pp. 76–100

Chapter 5
Textbook Approval System and Geography Textbooks in Japan

Minori Yuda

Abstract In this chapter, the author describes procedures of the textbook approval system and geography textbooks used in Japan's schools. Geography content within textbooks and skills required at each grade level are further explained to provide a picture of geography education in the country.

Keywords Geography education • Geography textbooks in school education • Japanese education • Textbook approval system

5.1 Introduction

To understand education in a country, textbooks can be an indicator and a data source for research. In Japan, most textbooks are written in Japanese. This language barrier may deter foreign scholars interested in the country's geography education. The author argues that it is worthwhile to implement a textbook system and textbooks in Japan that are in English.

The aim of this chapter is to provide information about the textbook approval system in Japan and to summarize textbook content to show the landscape of geography education in schools. First, the author explains the textbook approval system for school education and then introduce contents and skills required at each school level that meet the National Curriculum Standards.

In Japan, geography education starts from Social Studies in third grade and continues to the senior high school level. Here, the describes textbooks used from third to sixth grades in elementary schools and in junior and senior high schools.

M. Yuda (✉)
The University of Tokyo, 7-3-1 Hongo, Bunkyo-ku, Tokyo 113-8656, Japan
e-mail: minori@csis.u-tokyo.ac.jp

Y. Ida et al. (eds.), *Geography Education in Japan*, International Perspectives in Geography 3, DOI 10.1007/978-4-431-54953-6_5

5.2 Textbook Approval System in Japan

5.2.1 History and Procedure

All textbooks used in Japanese schools are published by private publishers after passing a textbook examination set by the Ministry of Education, Culture, Sports, Science, and Technology (MEXT). This is the textbook approval system established by the School Education Law in 1947 (MOFA 2005).

The process of textbook development and approval involves the following steps: writing and editing, screening, approval, publishing, and distribution. First, a publisher gathers together academics and school teachers. They read the Japanese national curriculum and standards for textbook examination before beginning to write a textbook. Following this, teachers discuss content with other members and editors during the writing process. The publisher submits a sample to MEXT. The Textbook Approval Research Council examines the text according to the Textbook Examination Standards. At the same time, textbook research officers from MEXT cross-reference samples with published standards. If the textbook does not meet a standard, MEXT requests revisions. The publisher resubmits a revised version, at which time the Textbook Approval and Research Council decides whether to recommend the textbook as appropriate learning material. The final decision rests with MEXT, who issues the approval. The entire textbook approval process is lengthy; most textbook projects take at least 3 years from the formation of the writing committee to actual use of the book in classrooms (MOFA 2005).

Textbooks are printed and distributed by many publishers, but the final decision on which books to use depends on local boards of education for public schools. For private schools, this decision is made independently. The local education boards and private schools notify MEXT of their textbook choices and the number of students who will use them. Publishers are ordered by the Minister to print enough copies to fulfill the total orders from all parts of the country. Figure 5.1 illustrates the process and actors involved.

Textbooks are provided to all students at compulsory school levels (elementary and junior high schools) for free, but students at the senior high level buy their own textbooks.

5.2.2 National Curriculum Standards and Textbook Approval System

The education content in Japan is defined in the *Gakushu Shido Yoryo* (National Curriculum Standards). The National Curriculum Standards are the education standards defined by the government. General rules in the National Curicculum Standards at each school level state that teachers must organize the education

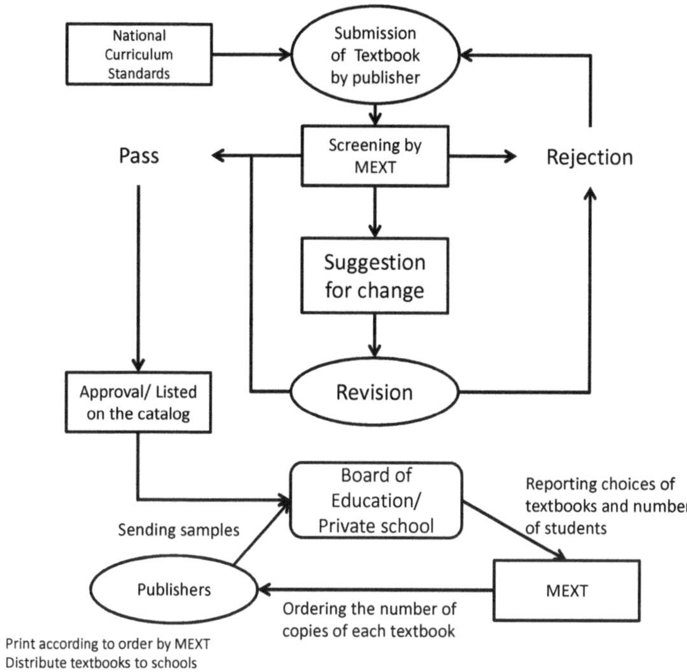

Fig. 5.1 Textbook approval system in Japan

curriculum in accord with the National Curriculum Standards. The latest national curriculum was introduced in elementary schools in 2011, in junior high schools in 2012, and in senior high schools in 2013.

The National Curriculum Standards are issued 3 or 4 years before introduction into schools. Soon after these curricula are published, publishers produce new textbooks that align with the examination before the new curriculum is implemented.

After publishing an approved textbook, the publisher begins work on the next edition of the textbook and publishes it 4 years later. This cycle of publication is also set by law.

5.3 Geography Textbooks in Japan's School Education

This section examines elementary school geography textbooks from Tokyo Shoseki, a publisher with the largest market share in Social Studies textbooks. Another publisher targeting secondary schools is Teikoku Shoin, with a market share exceeding 90 % in atlases and the largest share in secondary level geography textbooks (Watanabe 2011).

5.3.1 *Textbook Formats*

At the elementary school level, geography is part of Social Studies. One set of Social Studies textbooks (two volumes) are used in the third and fourth grades. In the fifth and sixth grades, one set of textbooks (two volumes) are provided for each year. Social Studies (geography area) in junior high schools and Geography A and Geography B in senior high schools require one textbook each. Book size in elementary schools is A4-size wide (210 mm) and B5-size long (257 mm). Since B5 size is common in Japanese schools, there are many textbooks and handouts of this size. Geography textbooks for secondary schools are also B5 size (182×257 mm).

The Volume One Social Studies textbook for the third and fourth grades is 108 pages in length and Volume Two is 162 pages. The fifth grade Volume One is 100 pages, and Volume Two has 116 pages. The sixth grade textbooks are 152 pages for Volume One and only 80 pages for Volume Two. The textbook in junior high school is 280 pages, and those of Geography A and B are 214 and 342 pages, respectively (Table 5.1). These textbooks are neither too large nor too heavy for students to carry back and forth between school and home.

Figure 5.2 shows the covers of these textbooks. All books are printed in color and include numerous maps, figures, and pictures. In addition to Social Studies

Table 5.1 Textbook formats at each school level

School level	Elementary school						Junior high school	Senior high school	
Grade (Age)	3 (9)	4 (10)	5 (11)		6 (12)		1–3 (13–15)	1–3 (16–18)	
Subject	Social Studies						Social Studies (geography)	Geography A	Geography B
	Vol. 1	Vol. 2	Vol. 1	Vol. 2	Vol. 1	Vol. 2			
Size	210×257 mm						182×257 mm		
Page	108	162	100	116	152	80	280	214	342

Fig. 5.2 Textbooks for geography in primary and secondary education (from left: Kita et al. 2012c; Nakamura et al. 2012; Arai et al. 2013 and Katahira et al. 2013)

textbooks, supplementary reading material dedicated to local municipality and prefecture issues is provided to third and fourth grade students by the local government. All students receive an atlas at every school level.

5.3.2 *Textbooks for Primary Education*

Geography is included in Social Studies at the primary education level. Students study Geography-related units mainly in the third, fourth, and fifth grades. Some geographic elements, such as comparative studies between Japan and another country or leaning about global issues, also appear in sixth grade.

The National Curriculum Standards for elementary school sets common objectives for grades 3 through 6, including the use of maps, globes, and statistics, and development of skills for thinking about social phenomena, their characteristics and interactions, and for expression of student ideas and research results (MEXT 2008b).

The National Curriculum Standards explain learning outcomes at each grade level. All textbooks are designed to achieve these objectives and include learning elements in the curriculum standard.

In the third and fourth grades, students conduct research in their community and begin to have deeper interests in their local area and prefecture. The term "local area" in this chapter refers to city and township levels. Fifth grade students will have a wider view of the entire nation, and students begin to learn and think about the entire world in sixth grade.

Students in the third and fourth grades conduct field surveys to learn more about the area surrounding their school and the town or city that they inhabit. They also become familiar with businesses and industries in their community by speaking to workers. Elementary students acquire basic knowledge of disasters, disaster prevention, and various systems (urban, water, and waste disposal). Students also learn about their prefecture through research and data collection. Table 5.2 summarizes content covered at the third and fourth grade levels.

Most teachers cannot use the third and fourth grade social studies textbooks for local studies, because these books use examples taken from select cities and prefectures. Since teachers use their own local community in class, they must develop similar content for their own areas. Teachers use textbook examples as a model for student activities and expected outcomes.

In thc third and fourth grades, the teaching method is for students to learn from experiences such as field surveys. Therefore, skills in these grades emphasize surveying methods, including interviews. Many types of map-reading skills and graphicacy are included in a section called Interpretation Skills. These skills include how to read maps, graphs, and tables. Map reading includes reading contour lines, topographic relief, and land use. More generally, use of an atlas is also taught. Therefore, textbooks are designed to include appropriate examples for local studies as well as some important geographic skills for students to acquire and use (Fig. 5.3).

Table 5.2 Contents of Social Studies textbook used in third and fourth grades of elementary school

Chapter	Section	Points
My town, town for everybody	Around school	Field research
	Condition of the city	
Workers and our lives	Staff in a store	Characteristics of our community and people's lives
	Farmers' life	
Changing life	Antique tools and old life	Change of people's lives
	Tradition to posterity	Learn history in local community
Protection of our life	Protection from fire	Disaster prevention education
	Protection from accidents and incidents	
To our better life	Where does water come from?	Urban system knowledge
	Disposal and reuse of waste	
Development of country	Irrigation network(s) in the mountains	Local history
Our prefecture	Location of prefecture	Characteristics of the prefecture
	Regional characteristics and people's lives	
	Prefecture connecting to the world	

Kita et al. (2012a, b)

Fig. 5.3 Pages from Social Studies textbook for third and fourth grade (section entitled "Around school") (Kita et al. 2012b)

In the fifth grade, students learn about Japan including its natural environment (e.g., landforms and climates), industries (e.g., agriculture, fishery, and manufacturing), and acquire information about society's role in environmental conservation (Table 5.3). The most common teaching method is for students to read maps, graphs, and figures.

Table 5.3 Contents of Social Studies textbook used in fifth grade of elementary school

Chapter	Section	Points
Our national land	Japan in the world	Understanding of natural environment in Japan
	Features of landscape and people's lives	
	Characteristics of climate and people's lives	
Our lives and food	Rice: Shonai plain	Agriculture and fisheries
	Fishery: Shizuoka	
	Food industry and our future	
Our lives and manufacturing industry	Automotive manufacture	Industry in Japan
	Manufacturing industry and industrial area	
	Manufacturing industry and trade	
Information society and our lives	Information industry and our lives	Information society
	Information changing society	Information literacy
	How to utilize information	
Our lives and environment	Our lives and forests	Environment conservation and disaster prevention
	We save the environment	
	Obviate natural disasters	

Kita et al. (2012c, d)

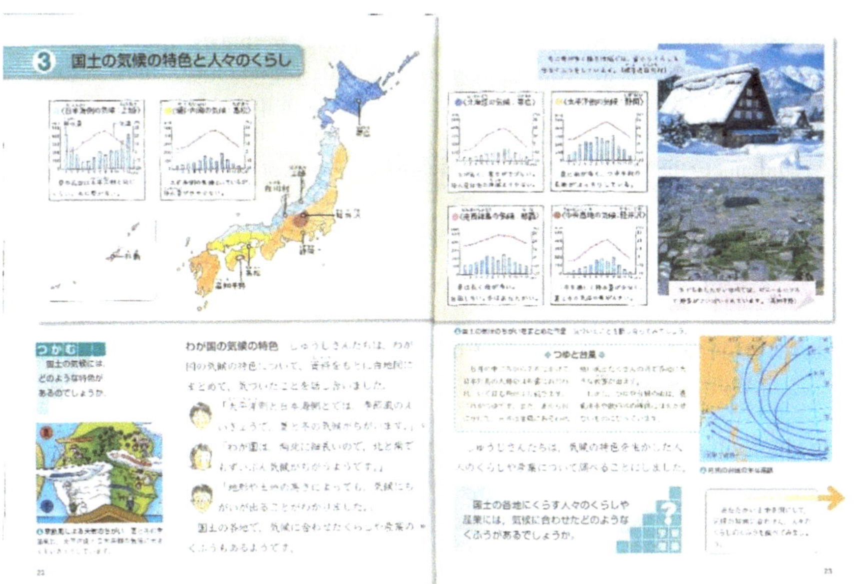

Fig. 5.4 Pages from Social Studies textbook for fifth grade (section entitled "Characteristics of climate and people's lives in Japan") (Kita et al. 2012d)

Many basic map-related skills are presented in the textbook for fifth grade students. Furthermore, use of the globe and interpretation of aerial photos are skills that students must acquire. These map reading skills underpin the next stage, in which students extract information from maps. Textbook content explains geographic facts, using photos and figures (Fig. 5.4).

Table 5.4 Contents of Social Studies textbook used in sixth grade of elementary school

Chapter	Section
1. History in Japan (History)	Jomon period to tumulus period/Building emperor-centered country/Generation of warriors/Muromachi culture/Warring states period to Edo period/Culture and new education in Edo/Meiji restoration/Japan toward the world/Protracted war and people's lives/To new Japan, peaceful Japan
2. Our lives and politics (Politics)	Politics make our wish come true
	Constitution of Japan and our lives
3. Japan in the world (Peace education and Geography)	Countries with ties to Japan U.S., Korea, China and Saudi Arabia
	Our future and role of Japan in the world

Kita et al. (2012e, f)

As students advance in grade level, required skills change from simply *reading* maps, graphs, and figures to *interpreting* maps or other figures and using them to inform their research or study. Equally important from the third through fifth grades are students' communication skills for arguing why their findings are important.

In the sixth grade, textbooks treat history from past to present and politics in Japan. The last section includes cultural geography of countries around the world (Table 5.4). This chapter includes peace education, since it is very important for students to think about how they can contribute to global society. This geographic content is very similar to what junior high school students study. This part may be good practice for sixth grade students to discover what they will learn in the next stage of education.

5.3.3 Geography Textbooks in Secondary School Education

5.3.3.1 Objectives of Geography Education and Textbooks in Junior High Schools

At the junior high school level, Geography is still a part of Social Studies. The main objective of Social Studies in the National Curriculum Standards is to help students "look wider and raise awareness of society, think from various perspectives using various data, understand and love our own nation and history, establish a basis for citizens to live in a global society, and make a peaceful and democratic nation and society" (MEXT 2008a, p. 31).

Table 5.5 shows the content of a textbook for junior high students. This book is normally used for 2 years. The geography curriculum in junior high school is designed for students to think about geographic phenomena from multiple perspectives. Students learn about and compare the world with Japan through their own research.

Geography in junior high school is focused on regional geography, so the curriculum emphasizes the understanding of differences and similarities in the world,

Table 5.5 Contents of a textbook of Social Studies (Geography area)

Chapter and sections
1. World Geography
– The appearance of the Earth
Continents, oceans, regional classification, major countries, longitude and latitude, globe and map, brief map
– Life in the world
Pacific Ocean islands, Siberia, Arabian Peninsula, Andes: houses, clothes, food and religion
– Natural environment, history and cultural background, industry
Asia, Europe, Africa, North America, South America, Oceania
– Research on another country
Decide on a country or region and a theme, collect source material, analyze data and make a presentation
2. Geography of Japan
– Appearance of Japan
Location, territory, prefectures, regions
– Comparison between Japan and the world: characteristics of natural environment
Mountains, plains, seashore and ocean current, natural disaster and prevention, population, natural resources and industry
– Geography in each region of Japan
Natural environment and energy in Kyushu, transportation in Chugoku/Shikoku, environmental issues and preservation in Kinki, industry in Chubu, cities and population in Kanto, life and culture in Tohoku, history in Hokkaido
– Research on an area
Set up a hypothesis, conduct fieldwork using topographic maps, summarize results and make a presentation

Nakamura et al. (2012)

as well as knowledge of and finding the uniqueness of Japan through learning about the natural environment, population, energy and relationships among regions. The National Curriculum Standards includes learning about the natural environment, history, industry, environmental protection, population and cities, life and culture, and relations between seven Japanese regions. The textbook focuses on one topic for each region.

The content of geography in junior high school mainly includes the explanation of locations and characteristics in the world and Japan. Therefore, teachers tend to give passive lessons focusing on rote memorization. This is the image of geography education in the country, which has often been criticized for this pedagogy. In response to these criticisms, many questions have been inserted into textbooks. These questions mainly ask students why a phenomenon is observed or occurs, revealing the most important way of thinking in geography education. For these reasons, most textbook pages are designed to include a substantial number of questions, maps, and graphs to support the development of student geographical thinking (Fig. 5.5).

Fig. 5.5 Pages from Social Studies (geography area) textbook for junior high school (section entitled "Characteristics of natural resources and industry: Energy production and consumption in Japan") (Nakamura et al. 2012)

5.3.3.2 Geography and Textbooks in Senior High School

In senior high education, geography belongs to the Geography and History subject category, and there are two subjects, Geography A and Geography B. Geography A is two credits, consisting of 70 lesson hours (two lesson hours per week). Geography B is four credits and 140 lesson hours (four lesson hours per week).

The common goals of these subjects are to consider various geographical issues in the world, to cultivate geographic points of view and ways of thinking, and to develop awareness for living responsibly in global society. The difference between these subjects is the approach to teaching. In Geography A, teachers impart characteristics of world regions, whereas Geography B teachers explain geographic phenomena using theories of systematic geography or by using regional geographic points of view.

The difference in these approaches is reflected in the content of each textbook. In the textbook of Geography A, many types of lifestyles across the world are explained from the viewpoint of landscape, climate, industry, and cultural differences. Many topics overlap with content in junior high school geography (Table 5.6).

The big difference between these two geography subjects is how phenomena observed in the world are explained. In the Geography A textbook, many pictures of

Table 5.6 Contents of Geography A textbook in senior high education

Part	Chapter	Section	Points
1. Appearance of the world and global issues	1. Contemporary society seen from globe and maps	Locations on the Earth and countries	Latitude and longitude, time zone, territory
		Globalized world	Relationship between Japan and other countries
	2. Environments around people's lives	Environmental factors to affect people's lives	In various regions in the world and Japan
		People's lives and landscape, climate, industry and culture	
	3. Lives and cultures in the world	China, Korea, Southeast Asia, India, Central Asia/West Asia/ North Africa, Sub-Sahara Africa, Europe, Russia, U.S., Latin America, Australia	Explaining dominant landscapes, climate, culture and industries in each region
	4. Global issues with us	Complicated global issues	Learning from examples from the world
		Environmental issues, resource and energy problem, population problem, food issues, urban problem	
2. Issues in local area	1. Various maps close to our lives	Maps and characteristics	Knowing maps and GIS
		GIS and technology for GIS	
		Utilization of GIS technology and maps	
		Creating different maps for different purposes	
	2. Natural environment and disaster prevention in Japan	Natural environment in Japan	Thinking of disasters and what we can do for disaster prevention
		Volcanos, earthquake, wind and flood damage and disaster prevention	
		Considering disaster prevention in the local area	
	3. Finding issues and area research	Deciding a topic, conducting research in local area and making a presentation	Nourishing research and communication skills

Arai et al. (2013)

landscapes and people are used for additional explanation, with brief text. In contrast, the Geography B textbook explains the phenomena and theories in depth, illustrated with pictures, maps, graphs and flowcharts (Fig. 5.6).

The Geography B lessons are designed for students to obtain a geographical way of thinking and to consider why different phenomena are seen. This departs from the traditional image of geography as a rote-learning subject. Students and teachers must adapt to the subject to develop geographical thinking and related skills.

The Geography B textbook begins with concepts of geographic information, GIS mapping software, longitude and latitude, and time zone difference (Table 5.7).

(Geography A) (Geography B)

Fig. 5.6 Pages from textbooks for Geography A and Geography B (section entitled "Landscape in Japan"): Geography B uses two pages to explain the same topic explained on one page in Geography A (Arai et al. 2013 and Katahira et al. 2013)

Table 5.7 Contents of Geography B textbook in senior high education

Part	Chapter	Section
1. Various maps and geographic skills	1. Geographic information and maps	Maps in contemporary world
		Types and usage of maps
		Mapping geographic data
	2. Utilization of maps and regional research	
2. Seeing contemporary issues from the viewpoint of systematic geography	1. Natural environment	Landforms in the world
		Climate in the world
		Characteristics of nature and lives in Japan
		Environmental issues
	2. Resource and industry	Development and change of industry
		Agriculture, fishery and forestry in the world
		Food problem
		World energy and mineral resources
		Resource and energy problems
		Industries in the world
		Tertiary industry
		Transportation and communication connecting to the world
		Trade and economic zone in contemporary society
	3. Population, rural and urban areas	World population
		Population issues
		Rural and urban areas
		Urban and settlement problems
	4. Lives, cultures, ethnic groups and religions	Life and cultures
		Ethnic groups and religions
		Nations in the contemporary world
		Ethnic and territorial problems

(continued)

Table 5.7 (continued)

Part	Chapter	Section
3. Geography in regions of the contemporary world	1. Regional classification in the contemporary world	Classifiation of regions
		Regional classification in different scales
	2. Regions in the contemporary world	Thinking from a regional geographical point of view
		East Asia, Southeast Asia, South Asia, West and Central Asia, North and Sub-Sahara Africa, Europe, Russia, Anglo-America, Latin America, Oceania
	3. Japan and the contemporary world	

Katahira et al. (2013)

Topics on systematic geography, phenomena and theories of landforms, climate, industries, population, rural and urban areas, culture, and religion are explained using figures and graphs. The section on regional geography begins by defining a theme, followed by explanation of characteristics using a regional geographic perspective for each classified region.

The National Curriculum Standards for senior high school, issued by MEXT in 2009, mentions the use of information and telecommunications networks. Specifically, GIS is mentioned as a tool to collect and analyze data using maps and spatial statistics. To comply with the curriculum standard, teachers must use GIS in geography classes. Hence, textbooks in both Geography A and B devote a couple of pages to GIS.

5.4 Conclusion

This chapter briefly introduces the textbook approval system in Japan, as well as content and skills in the latest geography textbooks from primary to senior high school education levels.

In Japan, all textbooks are developed by private publishers. MEXT screens textbooks to ensure that they are written in accordance with the National Curriculum Standards. After screening, only approved textbooks can be published and distributed free of charge to all students in elementary through junior high schools, and for sale to senior high school students.

In primary schools, geography education begins with a focus on the local community around the school. Textbooks deal with a range of topics, from gradually widening scales of city, prefecture, Japan, and the world. In secondary education, textbooks guide student thinking about Japan within the world, to interest them in various issues in the country and worldwide. These textbooks are aesthetically attractive and colorful, displaying fundamental skills of geography. The ideal of geography education can be found in these textbooks. Although their quality is not always matched by that in classroom teaching, good textbooks certainly contribute to effective class instruction.

References

Arai Y et al (2013) Kotogakko shin chiri A (Senior high school new geography A). Teikoku shoin, Tokyo

Katahira H et al (2013) Shinsho chiri B (New detail geography B). Teikoku shoin, Tokyo

Kita T, Sato M, Yoshida N et al (2012a) Atarashii shakai 3.4 ge (New social studies 3.4, vol 2). Tokyo shoseki, Tokyo

Kita T, Sato M, Yoshida N et al (2012b) Atarashii shakai 3.4 jo (New social studies 3.4, vol 1). Tokyo shoseki, Tokyo

Kita T, Sato M, Yoshida N et al (2012c) Atarashii shakai 5 ge (New social studies 5, vo. 2). Tokyo shoseki, Tokyo

Kita T, Sato M, Yoshida N et al (2012d) Atarashii shakai 5 jo (New social studies 5, vol 1). Tokyo shoseki, Tokyo

Kita T, Sato M, Yoshida N et al (2012e) Atarashii shakai 6 ge (New social studies 6, vol 2). Tokyo shoseki, Tokyo

Kita T, Sato M, Yoshida N et al (2012f) Atarashii shakai 6 jo (New social studies 6, vol 1). Tokyo shoseki, Tokyo

MEXT (2008a) Chugakko gakushu shido yoryo (National curriculum for junior high school). Higashiyama shobo, Kyoto

MEXT (2008b) Shogakko gakushu shido yoryo (National curriculum for elementary school). Tokyo shoseki, Tokyo

MEXT (2009) Koto gakko gakushu shido yoryo (National curriculum for senior high school). Higashiyama shobo, Kyoto

MOFA (Ministry of Foreign Affairs of Japan) (2005) Japan's school textbook examination procedure. http://www.mofa.go.jp/policy/education/textbooks/index.html, Accessed 28 Nov 2012

Nakamura K et al (2012) Shakaika chigakusei no chiri: Sekai no sugata to nihon no kokudo (Social studies, geography for junior high students: appearance of the world and territory of Japan). Teikoku shoin, Tokyo

Watanabe A (2011) 2012 nendo chugakko kyokasyo saitaku jokyo: Monkasho matome (Textbooks adopted in junior high school for the fiscal year 2012: summary by MEXT). Naigai kyoiku (Edu Inside Outside Jpn) 6126:6–8

Part II
Matters of Concern for Geography Education in Japan

Chapter 6
Reconsidering Regional Geography Education in Japan

Hiroaki Akimoto

Abstract The author reviewed regional geography education programs in Japan, with a focus on the National Curriculum Standards. The current curriculum has been constructed with the goal of helping students acquire knowledge about actual conditions in different regions in Japan and the world as a whole. The previous National Curriculum Standards, which was announced in 1998, was designed to center on ways of elucidating regional characteristics only. This change was criticized but the two approaches are essentially complementary. Regional geography education in Japan has some unique aspects compared to that in Europe and the United States, but shares the same objectives. Although the challenges of regional geography education lie in classroom instructional practices, issues within the education system itself, including the educational environment, also pose issues for geography education.

Keywords National Curriculum Standards • Regional and world knowledge • Regional geography education • Understanding regional characteristics

6.1 Introduction

The greatest change in recent revisions to the National Curriculum Standards for junior high schools and senior high schools in Japan is in regional geography education.

Regional geography education has played the role of providing school children with information about Japan and world regions, in addition to fostering an understanding of the world (Araki 2006). Since today's world is becoming rapidly globalized, knowledge about maps, locations, and places is important and thus raises the value of regional geography courses.

People now have access to large amounts of information about Japan and world regions through mass media and the Internet. Further, in the very volatile modern world, the situation in Japan and in the world is liable to change. This being the case, the meaning of geography as a subject will be called into question if its objective is merely to supply information about Japan and other world regions (Nishiwaki 1993).

H. Akimoto (✉)
Dokkyo University, 1-1, Gakuen-cho, Soka, Saitama 340-0042, Japan
e-mail: hakimoto@dokkyo.ac.jp

Y. Ida et al. (eds.), *Geography Education in Japan*, International Perspectives in Geography 3, DOI 10.1007/978-4-431-54953-6_6

So, how should regional geography education be organized? The recent changes can be seen as a response to globalization. This paper outlines recent revisions in regional geography education in Japan based on the National Curriculum Standards and suggestions on how regional geography education should be conducted in the future.

6.2 Topography, Regional Geography, and Regional Geography Education

Topography and regional geography are sometimes used synonymously, but a close examination reveals differences.

The oldest geographical record in Japan is said to be the *Fudoki*, records of geography and culture of Japan's provinces published in the eighth century. These were a set of comprehensive manuscripts detailing the situation in each region, created to help the rulers govern the land. In a democratic society, it is necessary to provide such information to the general public. Local government bodies produce prefectural directories, modern-day versions of *Fudoki*. Topography is a branch of regional geography, as discussed below, and refers to records made with the purpose of providing people with basic information about physical conditions in different regions.

The study of regional geography established in Japan in the nineteenth century. This is a branch of geography that aims to comprehensively study characteristics of a specific region. For successful study of such regional characteristics, activities such as observations, measurements, and interviews are carried out with regard to various phenomena in the region. Researchers examine the collected data to learn how phenomena are related and to make comparisons with other regions. Regional characteristics are clarified through analyses of this type.

Regional geography education is based on both regional geography and topography. However, differences between these two subfields are naturally reflected by different learning methods. Topography learning methods focus on providing information and understanding content. In contrast, regional geography places more importance on process, that is, perspectives, information gathering, and analysis. Shirai (2000) distinguished the two by calling the former "regional geography education" and the latter "regional geographic education". In this paper, however, the former is referred to as knowledge-focused regional geography education and the latter as methods-focused regional geography education. Needless to say, regional geography education has traditionally involved both topography and regional geography.

6.3 Role of Regional Geography Education

"Sixty percent of senior high school students cannot locate Miyazaki Prefecture". This was reported by Japanese newspapers in March 2008. This headline reflects the results of a survey conducted by the Commission of Geographical Education of the Association of Japanese Geographers (AJG-CGE 2008). Senior high school and university students took the survey. According to Sogo Shoto Kyoiku Kenkyujo

(Synthetic Research Institute of Elementary Education 2008), similar results were seen among elementary school students. Many regarded this as a serious problem, and their concerns were reflected in the 2008 Central Council for Education Report (Central Council for Education 2008). This report indicates the basic role that regional geography education is expected to play. It supports the view that students should acquire basic geographic knowledge about Japan and the world, both of which are necessary in today's society. Such basic geographic knowledge is certainly needed in daily life, but the meaning of geography in education goes beyond this. According to Gould and White (1974), humans create a kind of mental map and then use it as a basis for their decision-making processes. Thus, to develop in young people the capability to make decisions as members of a democratic society, it is essential to help students create mental maps that reflect an accurate understanding of regions and the world. This is the true role of regional geography education.

If instruction examines the local area, then students should be able to acquire geographic knowledge through direct experiences in their surrounding and it should be possible for them to develop regional understanding based on this underpinning knowledge. However, an understanding of Japan, other regions, and the world can only be built upon indirect information. Today, with the development of mass media and the Internet, it is very easy to access information about Japan and other places to help students form the basis of this understanding. However, information received through these channels is biased or incomplete. It is therefore essential that regional geography education in schools provide students with balanced information on the country and the rest of the world to helps them form an accurate understanding.

The understanding of regional geography education reflected above has also been advocated by Saito (2003), who asserts that regional geography education in schools has an important role in expanding student interest from the local "world" with which they are familiar, to the entire globe, which they cannot experience directly. Regional geography education also has a role in driving forward a change in perspective, from the animistic worldview of a young child to one governed by science.

In contrast, a position that places more importance on the formation of views and opinions than on acquiring knowledge has been around for some time. Yajima (1977) states that the aims of regional geography education are for students to correctly identify various geographic phenomena that develop on the earth's surface, and then to consider interactions between those phenomena and connections between humans and the environment. These experiences are to foster students' views and ways of thinking based on a regional perspective, and to cultivate a broad view of Japan and the rest of the world. In short, greater weight is placed on methods of forming views and ways of thinking, rather than on factual knowledge alone.

6.4 Problems and Solutions of Knowledge-Focused Regional Geography Education

Regional geography education in Japanese schools, especially elementary and junior high schools, has been carried out with the goal of providing information about Japan and other world regions, thereby fostering an understanding of the world. This is developed on the basis of knowledge-focused regional geography education. The approach involves dividing the world and Japan into regions and then imparting to students the conditions of each region. Japan is generally divided into seven regions (Hokkaido, Tohoku, Kanto, Chubu, Kinki, Chugoku-Shikoku, and Kyushu-Okinawa), and the world into Europe, Asia, North America, South America, Africa, and Oceania. The characteristics of each region are explained; however, there are a number of issues with this knowledge-focused regional geography educational approach.

To furnish students with a balanced knowledge-base of Japan and the world, teaching should deal with as many regions as possible (Shirai 2000). However, there is only limited available class time and, therefore, itemized, exhaustive rote learning rather than synthetic, deep learning is encouraged. This means that students often do not acquire an adequate understanding of the world. Such understanding does not only involve specific geographic knowledge, it also requires integrated and systematized knowledge. Thus, there is a need for geography instruction to be organized on the basis of concepts or methods representative of geographic characteristics in the region. Knowledge content has been developed based on the curriculum in Japan's regional geography courses, whereas methods content has had a supplemental role.

There are a number of highly regarded concepts in geography education, including general commonality and regional distinctiveness. In 1953, an article by Schaefer entitled "Exceptionalism in Geography: A Methodological Examination" criticized geographical work based on traditional empiricist descriptions. Schaefer argued that geography should seek to elucidate spatial laws and that regional research should be the testing ground (Tezuka 1988). Two years after Schaefer's article was published, the terms "general commonality" and "regional distinctiveness" appeared in the 1955 edition of the National Curriculum Standards for junior high schools in Japan (Ministry of Education, Science and Culture 1956). This was a pivotal move that quickly introduced contemporary trends of geographic thinking, already present in Europe and the U.S., to Japanese public education.

In regional geography education, the first method adopted to receive acceptance is called static geography. This method looks into as many of the geographic phenomena involved in a region as possible and provides explanations for them. Phenomena are then compared and contrasted across regions, and characteristics of each region are thereby ascertained. This is a reasonably effective method when considering traditional topographic descriptions and regional statistics, but it tends to promote itemized and exhaustive factual learning. Further, with limited time available for in-depth learning, it may not always be possible to extract the distinctiveness of a region.

The next method is dynamic geography. This method focuses on the phenomenon considered to be the most important in a region, then seeks to explain and link it with other phenomena. Individual phenomena can be discovered in any region, no matter how many or few. It is believed, however, that characteristics of the region are not seen predominantly through the phenomena themselves but rather in the ways that they are related; the way a region appears will be different depending on the phenomena studied. The consideration of which phenomenon to examine and how it relates to others is connected to deeper questions of how the structure of a region is perceived. This helps to cultivate a regional image.

A third method adopted under this topic is cluster sampling, also known as "learning by case study" or "sample study". During the study of a certain region, students focus on one part of it, analyze it in detail and are thus able to relate it to an overall understanding of the entire region. This is based on the idea that the local distinctiveness of a region extends across its various parts. The advantage of the cluster sampling method is that learning about relatively large-scale regions (e.g., nation states) makes it easier to grasp details of small-scale regions.

These methods were included in the discussion of ways to handle content in the 1969 National Curriculum Standards for junior high schools (Ministry of Education, Science and Culture 1969).

6.5 The Meaning of Methods-Focused Regional Geography Education and Related Issues

In the 1998 edition of the National Curriculum Standards for junior high schools (Ministry of Education, Science and Culture 1998), the focus on exhaustive regional geography education vanished and geography education was reorganized with importance placed on different scales, from communities and prefectures to countries and continents.

The 1998 National Curriculum Standards for junior high schools was based around methods-focused regional geography education. Criticism of this so-called cramming education and a rise in public demand for a more relaxed educational environment led to cutbacks in teaching hours and content. Regional geography education was required to follow suit.

Knowledge-focused regional geography education is based on the premise that every region has a unique existence. When a particular region is studied, the knowledge gained is isolated and not directly applicable to a generalizeable application across other regions. Under content-focused regional geography education, it becomes essential to deal directly with all regions. Reduced instruction hours in a knowledge-focused regional geography education system means a reduction in either the number of regions taught or the amount of learning content taught for each region. However, enlarging a regional unit would also increase the discrepancy between units within a course. Further, a reduction in content for each region would make multifaceted examination of regions very difficult, and instruction would be at a superficial level.

Methods-focused regional geography education has been adopted into modern topographic or area studies. According to Yagasaki (2003), researchers in area studies aim to understand regions by first setting out topics of study through exchanges and inquiries with people in the studied regions, before conducting fieldwork and related research. Yagasaki (2003) also argues that students who understand a region in this fashion will form a framework to comprehend the world. In other words, detailed study of a specific region is only the starting point to perceive the world in general. If this study method is introduced in regional geography education, schools would need to select regions of interest to students, then support their learning by simulating the experience of conducting original research in those regions. Understanding the world thus becomes a subject of lifelong learning. Because this method establishes students' interest as a starting point and leads to an individual understanding of the world developed through research and experience, it is well aligned with the constructivist approach to learning. This approach is also consistent with the recent trend in education of "learning how to learn". However, if this method is adopted, it will be essential to formulate an approach to regional geography education that is achievable through lifelong learning. Currently, the importance of such learning is not substantiated in detail within educational policy, although it is mentioned.

Methods-focused regional geography education revolves around activities wherein students conduct research and analysis to uncover the distinctiveness of a region. Teaching in line with these activities is more difficult than providing explanation about a region's distinctiveness. Every student has different interests in various regions. Therefore, teaching to students has limitations and detailed, attentive guidance is needed. Although digitization has provided students with tremendous access to materials via the Internet, there is a paucity of materials for student research and study. Thus, in methods-focused regional geography education, it is extremely important for the learning environment to be rich in such materials.

At a glance, methods-focused regional geography education is not truly effective at every stage of schooling. According to Shinohara (1984), there are four steps to cultivating an understanding of the world: (1) interest; (2) quiz-based knowledge; (3) scientific understanding and knowledge; and (4) subjective understanding founded on objective and scientific understanding (including understanding of oneself). Students at elementary and junor high school levels have little experience or knowledge of Japan and the world, so it is important that learning provokes their interest. There is also a simultaneous need to encourage practice and mastery of map reading at elementary and junior high school levels. Although knowledge-focused regional geography education is common in lower grades, it is absolutely essential to adopt methods-focused regional geography education techniques, such as surveys, analysis, and discussion, at the senior high school level.

Incidentally, Yamaguchi (1994) pointed out that although the recollection of place names, such as those of major countries, develops from elementary school until junior high school, knowledge of these locations are more developed in senior high school. The reasoning is that the knowledge of locations involves semantic and logical processing rather than mechanical rote learning, and therefore requires a higher stage of cognitive development. Some argue that from the perspective of establishing knowledge, regional geography instruction should use different methods in senior high schools from those in elementary and junior high schools.

In senior high school, both systematic and regional geography have an important role. From 1970 until the present, senior high school instruction about regional geography processes is more important than simply the knowledge of geography.

In junior high school, regional geography has been part of the curriculum for a long time. In 1998, the focus on regional geography processes was introduced into the National Curriculum Standards.

The 2008 edition of the National Curriculum Standards for junior high schools received criticism for its emphasis on methods-focused regional geography education, and the main focus reverted again to knowledge-focused regional geography (MEXT 2008). Knowledge-focused and methods-focused regional geography education are complementary, so it is important that both approaches are taught.

6.6 The Future of Regional Geography Education: Trends of Geography Education in Europe and the United States

Japanese regional geography education has considerable differences compared with that in Europe and the U.S. In particular, there is learning that fosters understanding of the physical world, including systematic geography, which occurs through the entire geography curriculum, beginning at the elementary school stage (Tabe et al. 2009). Regional geography education is positioned as a teaching unit that deals comprehensively with a specific region. In other words, regional geography education places importance on uncovering issues in a region and learning how we could respond to those issues.

This differs from the current position that regional geography occupies in the curriculum. In Japan, the main focus of geography education has been on the formation of a nation state and on conveying information about the country to its citizens. In contrast to Europe and the U.S., where geography developed as an academic discipline in which the natural and social sciences were combined for the purpose of developing, controlling, and administrating colonies. Therefore, geography education is not only concerned with citizen education but also with national territory. It is therefore recognized as an important academic field for the future of the state and its regions, laying the groundwork for national land-use planning, regional development, and environmental policy (Ikuta 2003).

Another difference is that spatial thinking continues to be a highly regarded concept within geography education in Europe and the U.S. According to the Geography Committee of the U.S. National Research Council, spatial thinking is applicable at three levels: daily life, reality, and intellectual (National Research Council 2006).

The daily life level is the most basic; spatial information at this level is gained through lived experiences. In geography education, learning one's way around the immediate community occurs at this level. The reality level is deeply connected to our daily lives but, at this spatial level, knowledge cannot be gained by direct experience. Instead, knowledge acquisition occurs through scientific methods. In geography education, study of a region that is not close to the learner takes place at this level. Learning at this level is a traditional function of geography as a subject (regional geography education). Finally, the intellectual level is a perspective from which to consider the future. At this level, cultural landscapes across the earth are considered projections of what humans of the past thought about, i.e., their intellectual space. According to this view, regions and the world as they will exist tomorrow must be conceived in the intellectual space of human beings living today. The argument is that the significance of geography education will increase with the implementation of teaching that goes beyond merely analyzing and understanding present conditions in regions, but that extends to the consideration of how oneself and one's world can and should be in the future. The background to this activity lies in the development of Geographic Information System (GIS) technology and information sharing.

It is from here that we begin to see the challenges of Japanese regional geography education. The first obstacle is how geography education should be imparted at the elementary school level. Geography lessons since the 1977 National Curriculum Standards (Ministry of Education, Science and Culture 1977) rarely included countries outside of Japan. A modest revival has been attempted since 1989, but the situation is still very different from that of Europe and the U.S., where children, right from the start of geography education, are encouraged to understand the world by comparing it with their home region and country.

The second issue concerns secondary school geography education. In Japan, the role of regional geography education in junior high school is to foster an understanding of the world. Regional geography education has been criticized for being too focused on rote learning. In response, a variety of schemes have been attempted that have accumulated a large amount of teaching experience. It would be pragmatic to make use of the accumulated lessons and to consider new approaches to effectively teach regional geography. Conversely, in senior high schools, while systematic geography education has a certain role, it is different from that of regional geography education in junior high schools. Japan should look to the manner in which regional geography education is conducted in Europe and the U.S., where importance is placed on uncovering challenges faced by a region and learning how they could be resolved.

The third issue concerns the practice of regional geography education and the direction in which it is moving. In Europe and the U.S., rather than keeping to a simple understanding of present conditions, regional geography education is clearly oriented toward the question of how future society should be constructed. This instills an active attitude for involvement in constructing regional and global civil societies. The importance of taking such a direction has also been indicated in Japan (e.g., Izumi 2009), and the National Curriculum Standards for junior high school (MEXT 2008) has clearly suggested this with its reference to investigating the challenges in a region. However, this statement does not appear to apply to specific educational activities that use materials such as data and maps. This is because of a lack of data and materials for student use, as well as little regard for the importance of these skills. Thus, teachers turn to case studies to illustrate regional phenomena. The more specific case studies become, the more students and teachers are likely to make value judgments and political appraisals of the regions studied. In the Japanese educational system, where political neutrality is considered important, education tends to go no further than teaching general concepts.

The central challenges of regional geography education in Japan lie in the design of curricula and onsite teaching methods. It seems this is not an issue of concern for educators in schools. For this reason, there is a real need to examine the education system, given the challenges of structuring geography education from elementary to senior high school and of preparing an effective educational environment.

References

AJG-CGE (2008) Daigakusei, Koukousei no chiriteki-ninshiki no Chousa houkoku (A survey report on the understanding of geography among university and senior high school students). http://www.ajg.or.jp/chirikyouiku20080319.pdf. Accessed 8 Dec 2013

Araki H (2006) Chiri tte nani ? (What is geography?). In: Araki H, Kawada T, Nishioka N (eds) Shogakusei ni oshieru chiri. Nakanishiya shuppan, Kyoto

Central Council for Education (2008) Shogakko, Chuugakko, Kotogakko oyobi Tokubetsu-siengakko no Gakushu Shido Yoryou no kaizen ni tsuite (Toshin) (Report concerning the reform of the designated the national curriculum standards for elementary schools, junior high schools, senior high schools and special support schools). http://www.mext.go.jp/a_menu/shotou/new-cs/news/20080117.pdf. Accessed 8 Dec 2013

Gould P, White R (1974) Mental maps. Penguin, London

Ikuta M (2003) Chiiki kenkyu no hatten to Chirigaku (The development of regional studies and geography education). In: Murayama Y (ed) Chiiki-kenkyu (Regional study). Asakura shoten, Tokyo, pp 1–24

Izumi T (2009) Chikyu-shimin no ikusei to Chiri kyouiku (The raising of world citizens and geography education). In: Nakamura K, Takahashi N, Taniuchi T, Inui T (eds) Chirikyouiku kouza dai1kan Chiri kyoiku no mokuteki to yakuwari. Kokon shoin, Tokyo, pp 191–211

MEXT (2008) Chugakko gakushu shido yoryo (National curriculum standards for junior high school). Higashiyama shobo, Kyoto

Ministry of Education, Science and Culture (1956) Chugakko gakushu shido yoryo shakai hen (National curriculum standards for junior high school, Social Studies). Futaba, Tokyo

Ministry of Education, Science and Culture (1969) Chugakko gakushu shido yoryo (National curriculum standards for junior high school). Okurasho insatukyoku, Tokyo

Ministry of Education, Science and Culture (1977) Shogakko gakushu shido yoryo (National curriculum standards for elementary school). Okurasyo insatukyoku, Tokyo

Ministry of Education, Science and Culture (1998) Chugakko gakushu shido yoryo (National curriculum standards for junior high school). Okurasho insatukyoku, Tokyo

National Research Council of the National Academies (2006) Learning to think spatially. The National Academies Press, Washington, DC

Nishiwaki Y (1993) Chiri kyouikuron josestu. Ninomiya shoten, Tokyo

Saito T (2003) Hassei-ron teki Chiri kyoikuron-Piage riron no Chiri kyoiku-ron teki tenkai (Developmental geography education—the development of piagetian geography education). Kokon shoin, Tokyo

Shinohara A (1984) Chiri kyouiku no Honshitu to Tenkai (Nature and development of geography education). Meijitosho shuppan, Tokyo

Shirai T (2000) Koko chiri no shin Gakusuhu Shido Yoryo wo meguru kadai (Comment on the new the national curriculum standards on senior high school geography). In: Nihon chiri kyoiku gakkai (The Geographic Education Society of Japan) (ed) Shin Gakushu Shido Yoryo to Chiri kyoiku (Geography education under the new the national curriculum standards). Nihon chiri kyoiku gakkai, Tokyo, pp 2–6

Sogo Shoto Kyoiku Kenkyujo. (Synthetic Research Institute of Elementary Education) (2008) Syakai ni tuite no kisoteki chishiki no syutoku ni kansuru Chosa (Study concerning the acquisition of basic knowledge of social studies). Sogo Shoto Kyoiku Kenkyujo, Gifu

Tabe T, Shimura T, Yamamoto M, Murayama A (2009) Gaikoku ni okeru Chiri kyoiku (Geography education overseas). In: Nakamura K, Takahashi N, Taniuchi T, Inui T (eds) Chirikyouikukouza dai1kan Chiri kyoiku no mokuteki. Kokon shoin, Tokyo, pp 117–170

Tezuka A (1988) Chirigaku no kakushin to dento (Reform and tradition in geography). In: Nakamura K, Takahashi N, Taniuchi T (eds) Chirigaku heno Shotai. Kokon shoin, Tokyo, pp 168–191

Yagasaki N (2003) Chiiki kenkyu no kadai to shorai (Issues in regional studies and its future). In: Murayama Y (ed) Chikikenkyu (Regional study). Asakura shoten, Tokyo, pp 170–191

Yajima N (1977) Chishi gakushu no kihonmondai (Basic issues in regional geography). In: Yajima N, Inoki J, Yamaga S (eds) Gendai chiri kyouiku kouza dai3kan Chishi gakushu no kihonmondai (Modern geography education seminar, vol 3: fundamental problems of regional geography). Kokon shoin, Tokyo, pp 1–14

Yamaguchi Y (1994) Chiri ishiki hattatsu no jisshoteki kyumei (A study on development of geographical cognition and affection in the elementary and high school students). Gunma daigaku kyoikugakubu kiyo, jinbun-syakaikagaku hen (Annual report of Faculty of Education, Gunma University, Cultural science series) 43:121–152

Chapter 7
The Current State and Issues of Field Research in Japanese Geographic Education

Shunsuke Ike

Abstract The National Curriculum Standards emphasizes field research outside the classroom. Yet the number of teachers who actually engage their students in this activity has declined considerably because it takes too much time to prepare and plan. The author argues that two actions are needed to raise the implementation rate of field research in classrooms. The first approach is to be clear about the value of that research to learning, which cannot be replaced by classroom teaching. Second, prior to taking on a teaching job, teachers need to experience and learn the best practices to introduce, and to strengthen the link between field research and geography content.

Keywords Field research • Geographic characteristics • Geographical skills • Local area

7.1 Introduction

"Geography is a study of the outdoors," a statement espoused by my former teacher and written on a blackboard in my first geography class at senior high school. Geography is based on outdoor activities and field research, although research topics are diverse.[1] Geography is a field science that values outdoor research activities as a form of investigation. Therefore, undergraduate and graduate education has invested substantial time and energy in the cultivation of skills necessary for conducting field research outside the classroom. The Association of Japanese Geographers established the accreditation system of "Field Researcher" and "Field Research Specialist," endeavoring to train specialists who engage in field research.

---

[1] Inquiry survey often occupies an important place in human geography, whereas observation is highly valued in physical geography. There are many levels of inquiry survey, ranging from simple research in which investigator collect data according to the fixed questions to advance research with a task of comparative culture, requiring researchers to view themselves objectively as represented in Ohno's study (1974).

S. Ike (✉)
Waseda University, 1-6-1 Nishiwaseda, Shinjuku, Tokyo 169-8050, Japan
e-mail: ike@waseda.jp

Y. Ida et al. (eds.), *Geography Education in Japan*, International Perspectives in Geography 3, DOI 10.1007/978-4-431-54953-6_7

The importance of field research is demonstrated by its inclusion in the National Curriculum Standards, which stipulates its implementation in geographical education from elementary to senior high school. The curriculum emphasizes field research, with mastery of conducting field investigation. Despite its inclusion in the national guidelines, Shinohara (2001) pointed out that field research is not popular in schools. Teachers of Geography and History, and Social Studies fail to fully understand the importance of field research. In this paper, the author presents the current landscape of field research from elementary to junior high and high school levels. In the process, tasks for outdoor field research are examined, its significance in schools explained, and suggestions are made to revitalize its favor in schools.

Although field research involves a variety of indoor activities such as literature research, statistical analysis, and the collection and analysis of maps, the core of this research is fieldwork, characterized by observations that include all sensory input. Thus, the term "field research" tends to be used synonymously with fieldwork. Distinguishing the two, however, the author focuses here on field research, especially on fieldwork. In the world of geography education, fieldwork tends to include activities such as field trips and excursions,[2] in which a director or guide plays a key role (Inui 2009). Therefore, this paper also groups these activities as fieldwork.

7.2 The Reality of Field Research in Elementary, Junior High, and Senior High Schools

7.2.1 Elementary School

Among the Social Studies classes taught in elementary school, field research was ranked the most important within community study for third and fourth grade students. The new National Curriculum Standards ask students to "observe and investigate social events through the study of a local area, a municipality and a prefecture, to cultivate the ability to use maps and a variety of concrete materials effectively". This includes thinking about the characteristics of social events within a community and their associations, and communicating the results of students' independent investigations (MEXT 2008a). Together, these indicate that implementation of field research is strongly called for in community study.

A relatively large number of elementary schools conduct field research in the third and fourth grades. For instance, a 1992 questionnaire survey of third and fourth grade teachers across 152 elementary schools in Kagawa Prefecture found that field research, on average, took place more than 3.5 times annually, and only four schools (3 %) failed to conduct any field research (Shinohara 2001). *Machi-tanken* (town expedition), which studies a community or municipality, was the most popular type

[2] Excursion refers to an activity in which participants visit a local place with a tour guide to observe its geographic phenomena.

of field research. Students observe and investigate topography and land use in their local areas, and there are opportunities for them to note local differences in the state of a district (geographic characteristics). In another study, a survey showed that 95.9 % of third-grade teachers in the city of Yokohama conducted field research in social studies (Matsumura 2011). The rate of field research implementation tends to be higher in elementary schools than at higher levels, because teachers in the former schools have more discretion to change the curriculum. Given this flexibility, they can decide to allot many uninterrupted hours to Social Studies.

Nevertheless, according to Shinohara (2000), many teachers believe that they fail to meet children's' demands for and keen interest in outdoor activities, owing to their own lack of skills. Many teachers are at a loss at how to teach geographic characteristics through field research. Thus, they often rely on Social Studies supplementary reading material prepared for the *Machi-tanken* class by each municipality. These reading materials, which put emphasis on learning development, have been popular recently. The demand comes from teachers who have insufficient content and pedagogical knowledge about a local area, especially for instruction in field research (Ike 2008). These teachers often lead outdoor activities with unclear goals and teach only superficial skills, such as how to take a compass bearing and recognize map symbols. Some teachers appear to shun the field research needed to understand the state of a district. This causes concern that field research, although included in the National Curriculum Standards, is not followed in the classroom.

7.2.2 Junior High School

Social Studies in junior high school consistently stress the implementation of field research, at the least the study of a local area. The latest National Curriculum Standards also set a goal to "enhance students' interest in geographical phenomena through concrete activities such as field research and to develop abilities and attitudes that enable them to think from various perspectives, judge fairly geographical phenomena by selecting and making use of various materials and express them properly" (MEXT 2008b, p. 29). In particular, the "investigation of a local area" primarily aims to deepen student understanding and interest in a local area through observation and investigation, and to help them master the perspective and method for investigating an area on the scale of municipalities (MEXT 2008b, pp. 66–67).

There are few teachers, however, who actually conduct field research even though it is expected that students actively implement it during their investigation of a local area. Empirical support comes from a questionnaire that surveyed 67 public junior high schools (128 teachers) in Miyagi Prefecture. Approximately 75 % of teachers answered that they only teach students how to read a map without going outdoors, and only about one-third of teachers lead students in conducting field research (Miyamoto 2009). This survey revealed that the major reason teachers failed to undertake field research was that they did not have enough time to prepare for and implement that research. This finding highlights the reality of teaching in junior high school. Teachers simply do not have enough time to improve the quality of course instruction because they spend their time instructing extracurricular activities, addressing students' behavior outside school, and executing other school responsibilities.[3] In examination papers, it is unlikely that senior high school students are tested on investigations of a local area; instead, they answer questions about reading a topographic map. As a result, most Social Studies teachers assume that students would not have trouble answering questions about local area investigation if they master map reading. This assumption is incorrect and contributes to the low implementation rate of field research.

7.2.3 Senior High School

The National Curriculum Standards emphasize the implementation of outside fieldwork, not only in elementary and junior high schools but also in senior high schools. The curriculum also expects that "students discover geographical phenomena by themselves, set an agenda, and carry out an investigation by adjusting research methods," using "geographical tasks as outlined in Geography A (MEXT 2010). For Geography B, it also states in "various maps and geographical skills" that students "must master geographical skills that conduct research on areas in which they can investigate directly, using maps from various perspectives and capturing geographical characteristics of living space" (MEXT 2010, p. 101).

Nonetheless, the implementation rate of field research is unexpectedly low even in senior high schools, despite the expectation that students at this level actively engage in field research. Low student participation was recorded in a 1990 survey, with responses from 66 geography teachers at public senior high schools in Ehime Prefecture. The findings indicate that field research was minimally implemented, if at all. However, 29 teachers (44 %) said that they introduced field research in their classes (Shinohara 2001). Field research in the prefecture was conducted in schools from 1963 to 1981, when geography was a compulsory or compulsory elective subject. Since then, however, many geography teachers have stopped conducting field research because they became very busy with a heavy schedule of school events, and student guidance and career counseling (Shinohara 2001). The most important factor causing the decline of field research was the difficulty teachers have had incorporating it into the school curriculum since geography became an elective subject. Therefore, the overall rate of field research implementation (including excursions) is low, although some geography teachers are dedicated to continue field research, as they believe that it reaps clear benefits.

[3]According to a questionnaire survey of 67 junior high schools in Kagawa Prefecture in 1992, only 15 schools carried out field research, and many merely asked students to read topographic maps (Shinohara 2001).

7.2.4 Causes for the Unpopularity of Field Research

Inui (1990) pointed out seven factors why the field research implementation rate was low: (1) Schools were increasingly concerned with accident prevention; (2) there were obstacles from specific environment factors around the school; (3) a lack of time for research implementation, because of an overcrowded curriculum and time allocated to university entrance examination preparation; (4) geography became an elective subject; (5) the length of time required to prepare for field research; (6) lack of a budget for travel expenses; and (7) a lack of teacher leadership. These factors vary by school. However, the fact that elementary school teachers are not sufficiently acquainted with the knowledge and methods of field research, and that junior high and senior high school teachers are too busy to arrange sufficient time for preparation and implementation of such research, are considered key reasons for a paucity of research in schools.

A number of factors explain teachers' failure to conduct field research. There is low teacher motivation for its implementation, although other factors such as institutional problems and lack of a supportive environment for field research play a role. Therefore, the first thing we must do to enhance the implementation rate is to help teachers understand the significance of field research. That is, we must answer basic questions such as "why is it necessary to introduce field research in the geography class?" and "what are the educational effects of field research that cannot be acquired in an in-class lecture?" To this end, the author discusses the significance of field research in the study of geography in the next section.

7.3 Significance of Field Research

Many researchers have pointed out the significance of field research in geography education (e.g., Shinohara 2000; Takeuchi 2002; Akimoto 2003; Inui 2009). Based on these studies, here the author advocates five factors that represent the value of field research.

First, students are likely to learn new characteristics of a local area through field research, which may enhance their motivation to learn geography. In such research, the subject of investigation is a "local area," i.e., the students' daily living space. Students often recognize this area as an ordinary, familiar, and mundane place. This attitude is likely to diminish their drive to learn (Shimura 2009). However, many discoveries and tasks that may be missed in their daily lives are often hidden in their living areas. As Ida (2000) points out, children are really impressed to learn that occurrences in a distant country can also be found in their local region. Such an unexpected finding, uncovered through field survey, motivates students to learn more about geography. Thus, field research is usually required at the start of a geography course so that students discover via experiential learning.

Second, it is easy for students to master the ability to discern geographic characteristics through field research. As Ukita (1970) indicates, a micro region such as a local area has different geographic characteristics that are interconnected. Therefore, it is generally easy to determine connections between various phenomena in a micro region. For instance, children have difficulty understanding the process by which agricultural products are delivered from producers to consumers when they only learn about Japan's distribution system on a macro level. At a micro level such as a village, it is comparatively easy to investigate the overall flow of a farmer's distribution of agricultural products to a greengrocer through a wholesaler. In this way, it is easy to recognize connections between phenomena in a micro region. To reveal geographic characteristics, we must focus on connections between different elements in a region. On this point, it is easier for students to identify such connections, through field research, between regional elements in a micro region than in a macro region. This can also be a fruitful training ground for the observation of characteristics in a macro region.

The third significance of field research is that it can be a yardstick for comparing regions. The learning as well as the experience from field research and direct observations provide a benchmark for studying regions inside and outside Japan. For example, if students can picture the size of a 1-hectare farmland through field research, they can accurately imagine the size of farmlands in a foreign country. Knowledge and a comparative ability acquired from field research can also function effectively as a standard of comparison when students try to understand the culture and social structure of another region. In addition, these experiences contribute to students rethinking the value of a local area through observation from a larger perspective.

The fourth factor is that field research provides a means for students in a geography class to learn geographical skills. Field research gives students the opportunity to observe scenery, learn fundamental research skills (e.g., reading and drawing a map, interviewing skills), which are difficult to learn in the classroom. In the classroom, a typical lesson on topographic maps takes place on a desk. Generally, students read contour lines and decipher the meaning of symbols. Route finding, however, a practical skill in real life, is difficult to acquire in the classroom. Thus, field research offers an authentic environment for students to master practical these geography skills.

The fifth benefit of field research is that it compensates poor childhood experience. As Takeuchi (2002) reports, children's perceptions of the social world have become distorted because many do not have natural and social experiences across a variety of settings.[4] The mass media also highlights children's underdeveloped space perception, self-centered thinking patterns, and a narrow-minded sense of value. These patterns are influenced by the changing environment in which children live, such as the trend toward the nuclear family, popularization of computer games, and the collapse of regional society. Students have an opportunity to learn about diverse values and life experiences outside these indoor situations through field research where students are

[4] As Teramoto (2002) pointed out, children today draw maps much more poorly than children in the 1980s, showing that their lack of experience of the real world greatly affects their development of space perception.

compelled to communicate with people of a different generation and with adults in the local community. These experiences support their building of lifelong skills such as creating social and professional networks (Takeuchi 2010).

7.4 Future Challenges of Field Research

7.4.1 *Issues on Field Research Content*

As noted above, field research can provide students with a variety of learning opportunities not possible through classroom teaching. The author believes, however, that we must examine the following tasks carefully to effectively conduct field research in geography classes.

The first task is to examine the content and methods of field research and align them with the developmental stages of children. Empirical research on the development of individual abilities has been accumulated,[5] but aside from Inui (2009) there has been no study aimed at designing field research materials appropriate for students' developmental stages at elementary, junior high, and senior high school levels. Therefore, it is necessary to examine how best to engage field research that is appropriate to each developmental stage.

The second task is to identify ways to nurture students' abilities to research geographical issues. As mentioned above, an unexpected discovery about their local area enhances their enthusiasm for learning. However, most students lack sufficient knowledge and experience to investigate important local geographic phenomena. Therefore, we need to teach them a basic perspective for observing a region of interest carefully beforehand, so they can select a theme and discover geographic characteristics and tasks. In reality, however, most teachers fail to teach students research skills to capture geographic phenomena, contributing to thee failure to initiate field research.

It is important to help students develop a perspective and sense of the local area under study, by first examining pictures of the region (Shibusawa 1985). Excursions are also effective in developing students' geographic perspectives. These activities, which are generally conducted in university-level geography courses, are expected to have a positive learning effect on students when conducted before implementation of field research (Miyamoto 2009). It is urgent to introduce excursions in geography education at elementary, junior high, and senior high school levels, to help students foster perspectives in evaluating geographic phenomena.

The third task is to investigate the association between field research and other learning content. Shibusawa (1998) explains that geography can be learned by

[5] For instance, Toriumi (1990) revealed that first and second grade students could see geographic phenomena only fragmentally in their observation from a school roof, whereas third and fourth grade students could make comparisons between geographic phenomena and explain their spatial relationships. Fifth and sixth grade students begin to recognize and explain geographic characteristics by viewing the entire scene.

direct and indirect experiences. Given this, the author suggests that it is necessary to apply learning outcomes and experiences acquired from direct field research to other learning content (e.g., the study of various regions). For instance, local field research is a useful method for students to discover geographic characteristics of a region, which is considered a key goal of regional geography education. The basic skill of identifying regional characteristics can be used to study regions of Japan and the world. Thus, the learned skill of observing geographic characteristics through field research contributes to another performance outcome. It is hoped that field research will become a more important part of the regional geography curriculum and general education in Japan and worldwide.

It is also possible to apply the skills and knowledge developed from field research to the study of systematic geography in senior high school. For example, in the 1970s, students studying Geography A (first year of high school) in Tama Senior High School in Kanagawa Prefecture conducted field research every Sunday for 7 weeks. Weekday classes on outdoor field research were also given. The content of these classes included: Uenohara-shi (river terrace) and Kofu Basin (alluvial fan) in Yamanashi Prefecture during the first trimester (unit dedicated to the study of the natural environment); Miura Peninsula (suburban agriculture and fishery), Hadano Basin (inland industry) and Yokohama Waterfront District (industry) in the second trimester (unit dedicated to the study of agriculture and industry); central Tokyo (urban function); and the Musashino Upland (location of settlements) in the third trimester (unit dedicated to the study of city and rural settlement).[6] With so many topics covered, this is simply an illustration that field research is useful not only for regional geography but also for systematic geography and the knowledge and experiences gained from field research can be associated with other learning content (Ike 2012).

The study of field research tends to be based on a unit of learning (e.g., "The investigation of the local area" in junior high school or "Geographical problems in the living area and field research" in senior high school), which fails to provide sufficiently in-depth investigation into a topic. The casual and occasional implementation of field research is insufficient for students to benefit from its value. Therefore, it is desirable that field research be introduced at each stage of geographic learning as necessary.

7.4.2 Institutional Issues to Implement Field Research in Education

It is important to improve not only the field research content but also institutional support to reinvigorate field research. The main reason that field research is less often introduced is that many teachers are unsure how best to teach it, because they

[6] Every first-year student was obliged to participate in a 1-day field research event at least once a year. This excursion experience offered valuable information in the "Annual Report of Geography," which compiles findings of outdoor field research. See Ike (2012) for the contents of Tama Senior High School's Geography A class, which used field research.

did not learn proper methods during their teacher training. For instance, elementary school teachers usually graduate from teaching training college or school, but these programs do not offer a class that focuses on the significance and methods of field research for non-geography majors. A survey of every student enrolled in the elementary school teaching license course at teacher training colleges and schools indicated that that only 11 of 49 universities (approximately 22 %) included field research in their curriculum (Shinohara 2001). As Ida et al. (1992) have remarked, teachers who experienced field research in college are likely to include it in their classes because they understand the importance of outdoor research.[7] Therefore, the author considers it necessary to establish a requirement for pre-service teachers to learn the significance and methods of field research, by participating in compulsory teacher training on the topic.

It is also necessary to build a system whereby relevant materials and information about field research are shared among teachers. In particular, elementary and junior high school teachers spend much time preparing for field research (e.g., the development of teaching materials for local area study). With a way to share materials (such as maps and lesson plans) and best practices, teachers with little field research experience can more effectively introduce it to their Social Studies classes. However, because of the lack of such a system, new teachers are left to design their lessons in isolation, with little support and relevant materials.

Since 1989, when Geography (and Japanese History) were combined to form the course "Geography and History," it has become an elective subject. The number of students who study Geography has declined since World History became a compulsory subject in senior high school. Since then, it has become increasingly difficult to conduct field research. It is especially challenging for field research to be done in distant provinces, another contributor to the poor implementation of field research in senior high school. It is critical now to consider how to restore geography to a compulsory status, as one way to increase field research implementation.

7.5 Conclusion

In field research, geographic characteristics and tasks become clear to students through careful observation of the scenery and data collected. The goal of field research cannot be replaced by classroom instruction. Knowledge gained from

[7] Nevertheless, according to a questionnaire survey of junior high school teachers in Miyagi Prefecture (Miyamoto 2009), 19 of 128 teachers (15 %) participated in field research when they were college students, among which only 6 teachers (31 %) conducted outdoor field research in junior high school. This shows that the experience of field research in college does not necessarily lead to a high implementation rate in class. There are few teachers with college field research experience, so it might be difficult to generalize this result. However, we cannot rule out the possibility that it has become difficult to implement field research in junior high schools owing to a tightening of conditions for its implementation.

such direct research serves as a reliable benchmark to measure the influence of geography education on students' thinking in school and throughout their lives. Thus, geography teachers must introduce field research as a way to highlight geography's direct application and importance to students.

It is not realistic for teachers unfamiliar with the process and topic to follow models of field research presented in the National Curriculum Standards and textbooks. Nevertheless, a lack of prior experience and training further increases teachers' hesitation to include field research in geography classes. The author suggest that they do not need to implement a high level of field research from the beginning; rather, a variety of field research should be conducted, depending on student developmental stage and teacher educational ability. As Sakurai (1999) and Toida (2007) report, simple research is appropriate at the elementary stage, such as investigating the distribution of elementary geographic phenomena and compiling new information on a map to identify spatial relationships. The key is for teachers to introduce field research and be confident that the quality of data, experience, and lesson plans will improve over time.

The best place to study geography is outside the classroom. If teachers understand this through their own experience, students will also gradually recognize the value and attractiveness of field research. The author hope for the reintroduction of field research in geography classrooms in the very near future.

References

Akimoto H (2003) Yagai kansatsu to chosa (Outdoor observation and research). In: Murayama Y (ed) 21seiki-no chiri – atarashii chiri kyoiku (Geography in the 21st century – new geography education). Asakura Shoten, Tokyo, pp 117–122

Ida Y (2000) Ningen keisei ni okeru yagai kansatsu/chosa no igi – Daigaku ni okeru kyoshoku kamoku no jissen wo tooshite (The significance of outdoor observation and research on character formation: through the practice of teacher training courses in a university). Tsukuba Daigaku Kyoikugakukei Ronshu 25(1):71–81

Ida Y, Fujisaki A, Yoshida T (1992) Shoto kyoin yousei gakubu ni okeru mijika na chiiki no yagai chosa ni kansuru shido – Jyoetsu kyoiku daigaku no ba-ai (The guidance of fieldwork on the local area in the department for training students to be elementary school teachers: in the case of Joetsu University of Education). Shin Chiri (New Geogr) 40(2):36–48

Ike S (2008) Shichoson gappei ni tomonau shakaika fukudokuhon no kadai (Tasks of supplementary readers arising from a municipal merger). Waseda Daigaku Daigakuin Kyoikugaku Kenkyuuka Kiyo 18:1–14

Ike S (2012) Chiiki chosa wo ikashita chiri jyugyo – 1970 nendai no tama koukou no Jissen no bunseki (Geography lessons utilizing field research: analysis of the case of Tama high school in the 1970s). Waseda Daigaku Daigakuin Kyoikugaku Kenkyuuka Kiyo 22:1–15

Inui T (1990) Chiiki no atsukaikata – Yagaichousa jisshi no jittai to kongo no kadai (The way of handling a region: the reality of fieldwork implementation and its future tasks). Chirigaku Hyoron 63A:177–178

Inui T (2009) Yagai chosa no arikata to kadai (Methods and tasks of fieldwork). In: Nakamura K, Takahashi N, Taniuchi T, Inui T (eds) Chiri Kyoiku Koza, vol 2: Chiri Kyoiku no Houhou (Lecture of geography education, vol 2: method of geography education). Kokon shoin, Tokyo, pp 319–330

Matsumura S (2011). Shogakkou 3nensei "midika na chiiki"no gakushu no genjho to kadai –yokohama shiritsu shogakkou wo jirei to site (The current state and issues of the study of "a local area" in the third grade of elementary schools: the case study of Yokohama City), Nihon chiri kyoiku gakkai dai 61 kai taikai happhou youshi syu (Proceedings of the 61st general meeting of the geographic education society of Japan), p 26

MEXT (2008) Syogakkou gakushu shido youryo kaisetsu shakai hen (The instruction manual for national curriculum standards of elementary school: social studies). Toyo kan shuppan sha, Tokyo

MEXT (2010) Kouto gakkou gakushu shido youryo kaisetsu chiri rekishi hen (The instruction manual for national curriculum standards of junior high school: studies of geography and history). Kyoiku Shuppan, Tokyo

MEXT (Monbu kagaku sho) (2008b) Chugakkou gakushu shido youryo kaisetsu shakai hen (The instruction manual for national curriculum standards of junior high school: social studies). Nihon bunkyo shuppan, Tokyo

Miyamoto S (2009) Chugakkou syakaika chiri teki bunya no"midika na chiiki" ni kansuru kyoin no ishiki (The consciousness of teachers on the study of "local areas" in the geography education of junior high school). Shin Chiri (New Geogr) 57(3):1–13

Ohno M (1974) Firudo waku no shisou – Sabaku no nouminzo wo motomete (The philosophy of fieldwork: in search of the image of desert farmers). Tokyo Daigaku Shuppankai, Tokyo

Sakurai A (1999) Chiri kyoikugaku nyuumon (Introduction to geographic education). Kokon shoin, Tokyo

Shibusawa F (1985) Chugakkou shakaika ni okeru kougai gakushuu (A field trip on social studies in a junior high school). In: Asakura Ryutarou sensei taikan kinen kai (ed) Shakaika kyoiku to chiiki gakushu no kousou (The scheme of social studies education and community study). Meiji tosho shuppan, Tokyo, pp 229–238

Shibusawa F (1998) "Mijika na chiiki" wo ikashita nihon no shochiiki gakushu no jugyo kaizen (The improvement of classwork on the study of japan's regional geography using the study of the "local areas"). Shakaika Kyoiku Kenkyu 80:9–20

Shimura T (2009) "Komedokoro" Echigo heiya no henyo – Chuugakkou shakaika chiriteki bunya "mijika na chiiki no chousa" no jirei to shite (Transformation of the Echigo plain as a "rice-producing region": the case study of the "investigation of local areas") - Geography in Social Studies on a Junior High School). In: Tabe T, Tajiri S (eds) Daigakusei no tame no shakaika jyugyou jissen nouto (Notes for the practice of social studies classes for university students). Kazama shobo, Tokyo, pp 70–76

Shinohara S (2000) Chiri kyoiku ni okeru yagai chosa no jittai to sono saikouchiku heno teigen (The reality of fieldwork on geographic education and proposals for its restructuring). Shin Chiri (New Geogr) 47(3/4):131–141

Shinohara S (2001) Chiri yagai chosa no susume – Shou, chuu, kou, daigaku no jissen wo toshite (Encouragement for geographic fieldwork: through the practices in elementary, junior high, and high schools and a university). Kokon shoin, Tokyo

Takeuchi H (2002) Yagai chosa no arikata to kadai (Methods and tasks of fieldwork). In: Takeuchi H, Kagami M (eds) Mijika na chiiki no shirabekata (How to investigate a local area). Kokon shoin, Tokyo, pp 1–6

Takeuchi H (2010) Jyugyo zukuri ni ikiru jinbun, shakai kagaku no kiso gijyutsu – Chiiki chosa no shuho wo manabu (Basic techniques of human and social sciences for the use in a classroom: learning the method of field research). In: Tokyo Gakugei Daigaku Shakaika Kyoikugaku Kenkyuushitsu (ed) Shogakko shakaika kyoshi no senmonsei ikusei, kaitei ban (Professional training for elementary school teachers of social studies, revised). Kyoiku shuppan, Tokyo, pp 46–53

Teramoto K (2002) Chizu no gakuryoku (Academic achievement of geography). Meiji tosho shuppan, Tokyo

Toida K (2007) Firudo waku shidou no kadai (Tasks of fieldwork guidance). In: Kobayashi K (ed) Jissen chiri kyoiku no kadai – miryoku aru jugyo wo mezashite (Practical themes in geography education: toward an attractive teaching of geography). Nakanishiya shuppan, Kyoto, pp 222–236

Toriumi K (1990) Yagai kansatsuryoku, shashin dokkairyoku oyobi chikeizu no dokuzuryoku no hattatsu to ikusei (Development and training of skills of outdoor observation, capacity for understanding pictures, and ability to read topographic maps). Shin Chiri (New Geogr) 37(4):40–47

Ukita T (1970) Chirigaku ni okeru chiiki no sukeeru – Toku ni nogyo chirigaku ni okeru (The scale of regions on geography: especially on agricultural geography). Jinbun Chiri (Jpn J Hum Geogr) 22(4):33–47

Chapter 8
Maps, Globes, Cognitive Space and Geography Education in Japan

Kiyoshi Teramoto

Abstract Maps play an important role in the formation of children's cognitive space. Students can extend their knowledge of place names together with their intellectual curiosity using skills of visual representation. Maps and globes represent various scales, from local to global, and they contribute greatly to the formation of our basic geographic world. However, the current geography curriculum in Japan provides inadequate support for students learning about geographical principles and the world.

Keywords Childrens' cognitive space • Geography education • Map learning

8.1 Introduction

Maps are tools that support geography education from a scholastic point of view, or they can be considered the very content of geography education. To that extent, maps hold the key to the development of geography education.

Generally, maps used in geography education are flat. They impart information with a high level of academic rigor, as represented by topographic and other types of thematic maps. However, when we think about the state of geography education and maps for individual children, it should be obvious that it is ineffective to teach geography without addressing their cognitive space. For learners, understanding spatial phenomena is simultaneously related to formation of their cognitive space. The role of geography education is important in illustrating the practical and academic aspects of geography. However, it must be recognized that geography education has a role in developing spatial awareness in children, even before it cultivates their intellectual learning. Human awareness of unfamiliar spaces is an indispensable part of living on Earth; it contributes greatly to the formation of our world view, which affects how we make decisions. In that sense, the convergence of maps, globes, and cognitive space should be the window through which horizons of geography education expand.

The school education in Japan is an important part of a child's development. Without exception, geography education primarily plays a role in Social Studies

K. Teramoto (✉)
Tamagawa University, 6-1-1 Tamagawagakuen, Machida, Tokyo 194-8610, Japan
e-mail: ktera@edu.tamagawa.ac.jp

Y. Ida et al. (eds.), *Geography Education in Japan*, International Perspectives in Geography 3, DOI 10.1007/978-4-431-54953-6_8

classes and is strongly influenced by the National Curriculum Standards. Here, the author describes the current state of geography education and offer some thoughts toward improving it, using maps, globes, and cognitive space.

8.2 Map and Globe Skills

8.2.1 The Relationship Between Maps and Globes

In the visualization of space, maps and globes have a complementary relationship. Maps are usually flat, whereas globes are three-dimensional (spherical). However, if we see the surface of a globe as two-dimensional, then it becomes a map representing the earth. The inability to acculately project globes is a limitation that cannot be ignored, but for the most part maps accurately depict the Earth's surface.

Geography teachers cannot possess any real vision if there is no concern for how learners read, interpret, and make use of maps and globes to advance their studies and daily lives. Teachers who consider geography important have a wish to impart to students the ability to use maps and globes. In the United States, these are referred to as map and globe skills, which are recognized in both education and general society (Tabe 2009).

Contemporary society is filled with complex information that cannot be correctly deciphered without map and globe skills. For example, news of an earthquake disaster in the Tibetan Autonomous Region of western China or reports of a mining accident in Chile are best understood by engaging map and globe skills to create a mental image of the space. Through these skills, one can visualize where the event occurred, climate of the location, vegetation in the region, or the kind of lives that victims led prior to the natural hazard. In particular, a globe is an effective tool for conceptualizing the seasons based on spatial information (e.g., location in the Northern or Southern hemisphere relative to the Equator).

Maps and globes have many uses in daily life. For example, when one is searching the location of a conference building in a city center, a detailed map of the city is invaluable. Use of maps is common for drivers with a Global Positioning System (GPS). In essence, we use and study geography every day. Although we may not use globes daily, their value is apparent when the need arises. Without a globe, it would be more difficult to understand what it is like on the other side of the world, to have a culturally sensitive conversation with Japanese-Brazilians, to examine the global impact of climate change or ocean pollution, or follow routes taken by famous explorers.

In other words, map and globe skills are tools for survival in contemporary society and a core form of knowledge.

8.2.2 Maps and Globes as Skills

Next, the author discusses the relationship between spatial cognition, map, and globe skills.

There are cases in which maps are fictional, but for the most part maps and aerial photos are depictions of actual spaces. Maps are expressions of spaces that depict things according to fixed rules, the "language of maps," which differs somewhat from human cognitive space. Stations, railway lines, and major arterial routes are represented with somewhat exaggerated sizes. Aside from pictorial maps and those made for a particular purpose (e.g., advertising flyers for real estate), maps generally do not contain many exaggerations.

Conversely, human cognitive space is riddled with exaggeration and distortion of real space. We tend to exagerrate the size of a place we are familiar with, and we underestimate the distances between familiar locations while unfamiliar or alien places seem farther away. Applied to an international context, countries like North Korea, Iran, and Afghanistan are somehow more challenging to locate in one's own world view. Cognitive space is imprecise and a product of a psychological phenomenon, which starkly reveals differences in one's knowledge. Maps and globes take on the role of correcting inaccuracies in our cognitive space, bringing it closer to reality. To accurately picture a place in our mind means to have visited the place. If we can create a mental map of our immediate surroundings, we feel more secure.

If we can recognize a place from a certain characteristic or items that we perceive to be relevant to the surrounding places, then we can develop a more complete picture of that place (Teramoto and Ohnishi 2004). Human beings can probably preserve their inherent sense of stability by going further than simply viewing scenery and objects in their surroundings along their behavioral pathways from a flat, sideways perspective. Nevertheless, people today tend to commute daily over the same fixed route, by car or train. Humans can feel more secure in their surroundings when they can imagine their route and landmarks along the way. For example, a person's awareness of multiple routes to their workplace and the ability to access an internal survey map that provides an aerial perspective is the essence of a map-like cognitive space. What is popularly called "hometown awareness" is also related to acquiring this kind of survey map. In this sense, cognitive space that includes maps and globes has a personal application. If one can make a map out of the cognitive space in one's mind, it would probably have a different appearance than a flat map or globe. Compared with these, a cognitive map has disparities in the amount of information perceived. A map represents the same level or volume of information across an areal extent, but a map of cognitive space is less complete.

People's cognitive space evolves with time and our interaction with space. When Japanese people first saw world maps and globes, they must have asked themselves questions like, "Is Japan really such a small island country?" or "Are we really a small country between big countries?" or "Do we have four seasons because we're located at mid latitude?"

If we compare the extent of cognitive spaces of Japanese people from the Edo through Meiji periods with those of people today, the former would have an affinity to maps and globes. If one looks at the globe and world map used in Kaichi Elementary School in the city of Matsumoto in Nagano Prefecture, one of the oldest elementary schools in Japan, we can see how much the Japanese people of earlier times would have wished to have an accuate view of the world.

8.3 Children's Cognitive Space Formation and Maps

8.3.1 The Anchor-Point Hypothesis

The late Professor R. Golledge, former president of the Association of American Geographers and faculty member at the University of California, Santa Barbara, explained the developmental process of human cognitive space and ongoing formation of mental maps based on the anchor-point hypothesis. This was modeled after the way that nautical charts are made from stationary points, such as the location where a ship's anchor is lowered to the ocean floor, and how the behavioral paths between one's home, workplace (or school), and recreation sites (or stores) show gradual divergence (Fig. 8.1). This hypothesis is intended to explain the process by which one comes to know unfamiliar spaces through experience, and aims to explain the development process of a mental map.

When studying the unit "The Shape of Our City" in elementary Social Studies, students learn by confirming the city location on a map, its landforms, the name of any river flowing through it, and locations of major public facilities and supermarkets. From this process a mental map is formed. After studying the names of adjoining cities, tracing the lines of railroads and expressways, and learning place names outside the city, students eventually recognize that the city boundaries extend outwards. If students can remember experiencing places they have physically visited and associate them with places on the map, the anchor points are tied together and

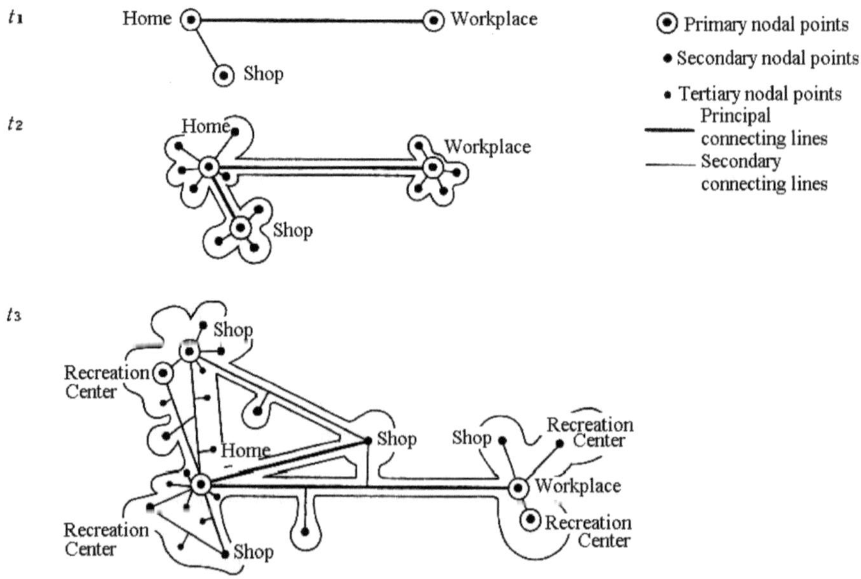

Fig. 8.1 Model diagram of Golledge's anchor-point hypothesis (adapted in part from Golledge 1978)

their awareness develops in a more complex way, such as, "Ah, I see, that place and this place have this kind of locational relationship." In other words, when students have visited a place and know about it, there is an associative process in the mind. Further, if it is a place they do not know, they make analogies from similar and already known areas; a so-called learning anchor point is formed that adds to mental map development (Teramoto and Ohnishi 2004).

In third grade Social Studies, learning key place names and identities of facilities provide anchor points, and surrounding second-order place names furnish supplemental information. For example, children who live in Tokyo's Shinjuku Ward use rail lines connecting locations of the Asakusa district (a famous tourist site), Chiyoda and Minato wards (where the Diet and other government agencies are clustered), and arenas in Saitama and Yokohama (sites of sports events and concerts). Along their travel routes, students can verify locational relationships between these places on a map. In the fourth grade unit "Our Tokyo Prefecture", student mental maps expand as they become aware of locational relationships while studying the silk weaving industrial zone of Hachioji or shapes of the seven islands of Izu. These examples are representative of how cognitive space is developed through geography education.

8.3.2 Necessary Instruction in the Period of Expansion in Geographic Awareness

The third and fourth grades are a period of expanding children's geographical awareness (Yamaguchi 2009). In the fourth grade, atlases are distributed and are important as engines to expand children's anchor-point knowledge about Japan and other countries.

Maps are composed by three kinds of symbols: point symbols that indicate a place on the earth surface (i.e., city hall or landmarks), line symbols that represent continuous landforms or structures on that surface (such as rivers or railway lines), and area symbols of fixed extent (such as land use or oceans). Printed on top of these are place names and lines of latitude and longitude. If these three map elements (points, lines, and areas) are included in the instructional methods then they contribute to students' cognitive space awareness.

Using these three elements in instruction, using a pointer or finger to indicate a point, trace a line, or outline an area, will certainly affect student learning (Teramoto 2002). When reading a map of Nagoya and its vicinity, the teacher can say, "You can find many point symbols for cars around Toyota City. Point out these symbols with your finger." In this way, students become aware of point symbols. Then, introduction of the line element may begin with, "Around Toyota City you can see many roads, along which cars are shipped to factories all over Japan and to foreign countries. With your finger, trace the lines of the main expressways and toll roads, as well as roads extending to the harbor." Students thereby become aware of line symbols. Next, "There are many factories not only in Toyota City, but all around Ise Bay.

Using your finger, outline this belt or area". This helps students develop an awareness of area symbols. Then a suggestion can be made based on such information, such as, "Imagine the extent of industries in the Chukyo Industrial Belt, home to the highest production in Japan." Random reading of map symbols does not contribute to the formation of a mental map. Rather, by adopting gestures of pointing to, tracing, and outlining points, lines, and areas with their fingers per the teacher's instruction, a coordinated feedback mechanism forms between the child's own body and the map information. Such experiences form a mental map in the student's mind and improve their level of understanding.

8.4 Maps in a Supporting Role Within Geography Education

Maps are useful learning material outside of geography education. They can serve as a language for conveying the locations of events, and can be applied across academic fields in the natural and Social Sciences to all aspects of our daily life. Maps are also important in geography education. However, if maps are more widely recognized as key to developing children's cognitive space, wouldn't it be more appropriate to assign maps a supporting role in geography? Geography education divides its content between topics of landforms and climate, industrial locations, regional change, trade, leisure activities, relationships between humans and nature, and a geographical view and way of thinking. Each of these topics relies on maps as a way to record and represent information. A variety of thematic maps have been created, with many maps and atlases sold in bookstores. It is necessary to recognize the supportive role of maps in geography education and to communicate their characteristics to the wider society.

The National Geographic Society (NGS) headquarters in Washington, D.C. showcases geography in a different light than in Japan. One reason for this is that the U.S. society is actively educating the public and schools about the value of geography in astronomy, biology, geology, history, and medicine. The NGS produces a variety of maps that engage people's intellectual curiosity about geography.

In Japan, there is a strong image of geography as belonging to Social Studies as a school subject, and the view of maps is tied to a fixed concept. Maps used in medicine show body regions or disease symptoms, and maps can also show the habitat of wild animals. However, these types of maps do not appear in geography education in Japan. If society understood the extent to which maps are used in different professions, geography education would be a leader in map use. Thus, maps should be elevated from a supporting actor to a substantial role in geography education.

It is regrettable that globes are not used more in our everyday lives. It is a waste for them to be used as nothing more than decoration in a company reception room or the teacher prep room in a Social Studies department office. If people developed the habit of looking at globes when thinking about world events, environmental problems or international conflicts could be better understood. Then, Japanese people would be more active in dialog about world events. To imagine how people are living right now in Paris or New York is part of an international mindset. Exercises in measuring distances or areas on a globe, verifying place names, and calculating time differences are all basic methods of globe reading. However, it is probably more important to make it a habit to simply gaze with curiosity at a globe.

8.5 Conclusion: Actions to Heighten the Social Usefulness of Geography

Maps and globes not only improve people's cognitive space, but they are also catalysts for change in one's image of the world and life. People who are adept at map reading are also competent in responding to spatial issues and can navigate with accuracy. For example, map reading heightens people's abilities to evacuate quickly during emergencies, decide on optimal locations for a business or home, make smooth and efficient journeys, quickly discover spatial relationships between multiple phenomena, and perform numerous other spatial activities.

There are three types of regions in geography: formal, homogeneous, and nodal. These are captured by by maps in different ways to communicate a theme. Some examples include: the gradient of rental cost change from city to suburbs; deciphering economics through theories such as Thünen's isolated state or Weber's industrial location; land-use analysis or criminal profiling using GIS. In the business world or as a tool for solving issues, geography education offers a means to educate people about the usefulness of maps.

What is needed in geography education is a textbook that communicates the importance and relevance of maps to social decision making. Within the Japanese context, display of themes such as disaster prevention and tourism on maps are becoming necessary tools. In urban areas where flash floods from localized heavy rain are becoming more frequent, if every resident developed a mental map of the area (the ability to mentally picture a precise location, elevation, and sense of distance), it would be invaluable to approximate the time to evaluate and reach escape routes. In such life-and-death situations, a vague spatial awareness is not adequate. It is only when this level of socially useful knowledge is integrated with an accurate sense of space and map-like awareness does it accumulate and become living and working knowledge. In that sense, the more information can be used to form a mental picture that is useful in society, the more valuable maps will be regarded.

Geography education is where the social usefulness of maps can be most enhanced, which is especially important because of the stereotypes people have about geography. The value of geography education is found in courses at the elementary, junior high and senior high school levels, as well as in general education. These subjects come under subject areas entitled Life Environment Studies, Social Studies, Geography and History, and Geography (including descriptive regional geography), plus Geography and History Education Methods. Maps are the principal tool of geography education and of its knowledge content. Maps are also the learning materials that connect set goals, methods, and learning content in teaching. Thus, the value of maps is not limited to learning about Social Studies and Geography (Shimura 2010).

Acknowledgments I wish to express my thanks to Shimura Takashi of Joetsu University of Education and Tabe Toshimitsu of Nihon Women's University, both of whom gave me many suggestions.

References

Golledge RG (1978) Learning about urban environments. In: Carlstein T, Parkes D, Thrift N (eds) Timing space and spacing time, vol 1. Edward Arnold, London, pp 76–98

Shimura T (2010) Shin gakushu shido yoryo no shushi wo fumaeta chizu gakushu no arikata (Status of map learning with regard to the new curriculum guidelines). Chizu (Map J Jpn Cartogr Assoc) 48(2):19–28

Tabe T (2009) Amerika shoto shakaika ni okeru chiri gakushu kojo puroguramu–monogatari kyozai aren sei saku ga, Sofu no tabi wo katsuyo shita jirei (The program to improve geography learning in beginning Social Studies in America– a case using Allen Say's Grandfather's Journey as teaching material). Shin Chiri (New Geogr) 55(3–4):19–32

Teramoto K (2002) Yomigaere chizu wo yomu chikara (Resurrecting map-reading skills). In: Teramoto K (eds) Shakaika no kiso kihon– chizu no gakuryoku (The fundamental basis of social studies– map reading skill). Meiji tosho, Tokyo, pp 9–64

Teramoto K, Ohnishi K (2004) Kodomo no shokokai– asobi kukan to tanken kodo no chirigaku (A child's first voyage–the geography of play space and exploratory behavior). Kokon shoin, Tokyo

Yamaguchi S (2009) Chiri shiso to chiri kyoikuron (Geographic thought and geography education theory). Gakubunsha, Tokyo

Chapter 9
Social Participation in the Community and Geography Education

Hirokazu Takeuchi

Abstract In this paper, the author considers the viewpoints of geography education practice and curriculum in the context of social participation within the community. Specifically, the author examines the significance of social participation learning in geography education. The author proposes that the geography curriculum endeavors to develop children's identity, which is complex and layered. The instruction mirrors the cognitive development by first teaching about hamlet, town, city, region, state, nation, and finally the world. This approach is one of nested multilayered regions on different spatial scales and it starts by building on children's direct experience in their local area.

Keywords Community • Formation of multilayered regions • Geography curriculum • Local area • Social participation learning

9.1 Introduction

In recent years, learning through social participation in school education has attracted public attention.[1] This arose from an item inserted into the Revised Basic Law on Education of 2006, Section 2, entitled "The Goals of Education," which reads: "Together by stressing justice and responsibility, equality of males and females, respect for oneself and others, and cooperation, to participate as actors in the formation of society based on public-spiritedness, and to nurture the attitudes that will contribute to that development." Subsequently, the School Education Law was revised in 2007, followed by the Revised National Curriculum Standards (MEXT 2008a), which also advocated fostering students' abilities and skills so that they could participate in public affairs. Among these trends, social participation is one focus of a geography course in junior high school entitled "Survey of one's local area". This course introduces the foundation of geographical perspectives

[1] Children's social participation can be at all spatial scales, but in this paper the main object for consideration is participation in the local area, which children can do through direct experience.

H. Takeuchi (✉)
Chiba University, 1-33Yayoi-cho, Inage-ku, Chiba 263-8522, Japan
e-mail: htakeuchi@faculty.chiba-u.jp

Y. Ida et al. (eds.), *Geography Education in Japan*, International Perspectives in Geography 3, DOI 10.1007/978-4-431-54953-6_9

and thinking, through experiential and project-based learning in the local area and students' living spaces. The curriculum simultaneously nurtures student abilities, skills, and attitudes for proactively participating in the development of their community.

The need for practical exercises for children that incorporate social participation was suggested earlier by geography researchers and educators. For example, Ida (2005, 2006) emphasized participation as a geographic skill within the learning process. Nishiwaki (2008) suggested a complete overhaul of the geography curriculum that consistently incorporates nurturing of student social participation abilities. The author also argued that to overcome the dwindling local area consciousness[2] that had arisen as a result of children's inadequate personal experiences (of both society and nature) in their local area in recent years, there was a need for instruction that includes local social participation (Takeuchi 2006).

This paper is based on these accounts, providing evidence to support the view that practical activities in geography education should incorporate social participation in the local area and be a model for the geography curriculum.[3]

9.2 Social Participation in Social Studies and Geography Education

9.2.1 The Formation of Awareness of Society and Fostering Citizenship[4]

The discipline of Social Studies, which was established in 1947 to spearhead post-war democratic education, states. "The mission of the newly established discipline of Social Studies from now on is to help our young people understand society and

[2] In an earlier paper (Takeuchi 2006), the author defined regional consciousness or awareness as a holistic mental response to children's local area, which includes aspects of conscious recognition and an emotional or sentimental one.

[3] Regarding adoption of a social participation perspective by the education bureaucracy in cases such as the Revised Basic Law on Education, Satō (2002) has indicated problems of "contradiction between self-initiated activity and mobilization in the emphasis on 'service' and 'love of homeland,' based on nationalistic views of public education and through linking together the school, the family, and the local area" (p. 201). In this paper, while giving due consideration to such problems, the author value the positive aspects of social participation learning that can foster active citizens through direct experience in the local area, and consider the value of such learning in instruction.

[4] In the National Curriculum Standards, the term "civic abilities" are used in the goals for the educational discipline. The National Curriculum Standards 1968 edition (Ministry of Education, Science and Culture 1969) is where civic abilities were first clearly spelled out in the general goals. The word citizen "should be understood as including two meanings, that of a citizen who is a member of civil society, and as a citizen who is a member of the nation" (p. 2). This generally supposes that civic abilities are those of individuals who are both local and national citizens. In this paper, to assume social participation in the local area and stress the attributes of the local citizen within "civic abilities", the author use the term "citizenship".

its livelihood and to foster the attitudes and abilities in them to contribute their energy to its development." Thus, its purpose was stipulated to help young people understand society and foster attitudes and abilities for contributing to building society (Ministry of Education, Science and Culture 1947). The nature of Social Studies is to provide students with the knowledge and understanding of society necessary to become contributing citizens. It was defined as a discipline that "fosters citizenship through an awareness of society" (Itō 1971) and Social Studies continues to play this role.

Geography education, which is part of Social Studies, contributes equally to citizenship formation. For example, the study of one's immediate local area or *Kyōdo* (native place) in junior high level geography, has two aspects: (1) learning about geographical knowledge, techniques, and geographical perspectives as well as ways of thinking; and (2) nurturing students' interest in their local area (*Kyōdo*), which leads to a desire to contribute to its advancement (Kikuchi 1969).

Outcomes from these two aspects, as Shibusawa (1998) points out, have been included in the National Curriculum Standards since 1977. Thus, an arguably important part of fostering practical citizenship is learning from social participation, the central topic of this chapter.

9.2.2 Social Participation in the Community

Within Social Studies and geography education, the relationship between the development of social perspectives and citizenship has been changing over time.

In the early inception of Social Studies, a large part of the curriculum emphasized child participation in their local society. Social Studies in the mid-twentieth century, such as the Nishi-Tama Plan (Imai 1950), supported children's identification of problems in their community and the search for solutions. Together with adults, students were actively involved in improving the local area.

However, the approach to learning through problem solving in early Social Studies was criticized. From the mid-1950s onward, there was a shift toward systematic learning, and the social participation element in Social Studies and geography was gradually phased out. For example, the 1958 National Curriculum Standards (Ministry of Education, Science and Culture 1958) included the idea of *Kyōdo* in junior high geography, which was a shift in pedagogy centering on methods and concepts (Kikuchi 1960). Specifically, in the *Kyōdo* lessons, the goals advocated the inclusion of uniform targets and methodological concepts. However, implementation into lessons was difficult. Instead, the curriculum focused on guiding geographic perspectives and thinking, as well as geographic techniques (e.g., map reading and construction) through local surveys and field trips followed by methods of analysis. Thereafter, lessons about one's local area or *Kyōdo* in Social Studies and geography focused on learning methodological concepts instead.

When Japan entered a period of rapid economic growth between 1954 and 1973, problems such as depopulation, overpopulation, and environmental pollution posed serious concerns to society. In this period, lessons became focused on teaching about issues rooted in the local area, with community problems used as teaching materials.[5] For example, a teacher and students in a polluted area of Fukushima (1981) used a number of data collection methods (i.e., questionnaire, interview survey, and chemical analysis) to research the extent of environmental damage in the community. They presented the results to local residents, creating important materials to inform decisions about the future of local development.

In the early 1980s, as social problems like environmental pollution declined, lessons directly involving student participation in community problems were reduced. Given the advancement of globalization, lessons adapted to social problems at the global scale, including environmental and development issues. The slogan "Think globally, act locally" encouraged action in the local area where students resided, with a global perspective (Ministry of Education, Science and Culture 1991).[6] In geography education, environmental and development issues were made into study materials, and classes comparing global with community problems were created. However, it was rare that such lessons produced any kind of social participation. Recently, attempts have been made to include social participation in the local area (Sudō 2006) in the "Period for Integrated Studies". However, in school classrooms today, where class hours have been reduced and schedules are crowded with competing demands, social participation in the class is an exception to the norm.

9.3 Need for Social Participation Learning

9.3.1 Significance of Social Participation Learning

Hart (2000) stated that along with developing their own awareness of society and socialization through all types of direct experiences, children learn about the principles of democracy, its realization as well inadequacies. Armed with this process of participation, children can grow into autonomous citizens within their community, with the knowledge to exercise their rights and responsibilities.

If Hart's assertions are applied to geography education, the significance of social participation learning in the community can be expressed by four points:

- Participation learning is a way for students to become aware of society (local area awareness). The study of geographic knowledge, developing geographic perspectives and thinking, and gaining proficiency in geographic techniques are effective through concrete and personal experiences in the local area.

---

[5] In this paper, the term "community problems" refers to environmental, development, poverty, income inequality, and human rights problems within the local area (assuming a spatial scale of the school district, where it is possible to directly experience things at the levels of city, town, or rural township).

[6] In this book, the Belgrade Charter (1975, p. 3) is introduced, which sets out six conditions for solving environmental problems, namely, concerns, knowledge, attitude, technology, evaluative ability, and participation. The book points out the importance of motivating behavior.

- Participation learning offers a venue for children to learn about citizenship. In the process of using community problems as the object of study, student interaction with adults and different exchanges will foster their abilities and skills to reach self-realization as autonomous citizens.
- By investigating the local area surrounding the school, students gain direct personal experience to their surrounding region. A lack of exposure to authentic learning is an obstacle for children's development, leading to adverse effects on their interpersonal relationships and social skills (Sumida 2001). Participation learning provides abundant experiences for students to learn about nature and society in their local area.
- Children's social participation learning may become a catalyst for local area improvement. This type of learning is supported by many different actors in the local area. As study becomes more advanced, a social network is naturally formed between students and adults who support their work.

9.3.2 Creation of Social Participation Learning in Schools

When creating social participation learning courses, educators should be mindful that while actions are public, value judgments and decision-making processes are personal (Takeuchi 2010). Teachers should recognize that social participation is voluntary. With voluntary participation being fundamental to participation learning, the following perspectives underpin the creation of materials and curriculum:

1. It is necessary to organize high-quality and meaningful experiences in the local area, such as survey activities and exchanges with different stakeholders in society.
2. Ultimately, educators must respect children's personal autonomy and independence, and create classes with a free atmosphere. Social participation should be self-motivated and that students voluntarily participate in their community.
3. Students should be encouraged to focus on deep personal reflection as they advance through the learning process. In a research study, the author analyzed the process of voluntary social participation activities of senior high school students. It became clear that when students encountered difficult situations, they overcame problems through a dialogue with themselves (Takeuchi and Sakuma 1996). In the context of social participation, deep personal reflection is important to problem solving, thus the value of time dedicated to this activity.
4. Comparison of community issues across different areas sheds light on similar problems in Japan and in the world. In other words, local issues should not be narrowly associated with a particular area, rather, there should be opportunities for students to discover that local problems have implications in the broader society.
5. Social participation is not simply an end point in geography education. Social participation engages students, who are encouraged to investigate solutions to societal problems and foster an ongoing dialog with oneself and with society.

9.4 Perspective of Social Participation Learning in Geography Education

9.4.1 Establishing Children's Identity in the Formation of Multilayered Regions

In exploring the concept of self-governing community-building by citizens from a sociology standpoint, Matsuno (2004) argues that the goal should be a citizen-initiated form of regional social operations. This idea positions the fundamental values as coming from a new "citizen-based communality" instead of the traditional top-down administrative communality, thus Matsuno coining activities such as "local area societal formation" (pp. 1–5). From another perspective, if Social Studies can fulfill the role of fostering students' ability to form a civil society, then the lack of critical thinking skills and value judgments in the current curriculum will be resolved (Ikeno 2001). This refers to a set of abilities and skills for creating a society based on democratic principles. The assumption is that self-sufficient citizens who make up a democratic society are those described in the National Curriculum Standards as "shapers of a peaceful and democratic nation and society who live within an international society" (MEXT 2008b).

Social participation learning aims to nurture children to become actors within a local society, with rights and responsibilities, i.e., citizens involved in "local area societal formation" (Matsuno 2004). The elements of social participation learning in geography education include abilities and skills that do not end at the limits of local area societal formation. Thus, such learning fosters Ikeno's aforementioned ability to form civil society. Geography is well positioned, because it imparts students with knowledge of geographic phenomena and social problems that occur on different spatial scales (local area, national scale, international and global society). The abilities and skills that students cultivate through social participation learning are at times learned from residents and citizens in their immediate local area, and at times from citizens nationwide or worldwide. The following section discusses children as stakeholders in shaping multilayered regions at varying spatial scales, aligning with a principle proposed in the national curriculum of "identity in the formation of multilayered regions."

The movement to establish this identity in children has attracted attention in the field of development education, which has aimed to cultivate practical global citizens. Yamanishi (2008) observed that practical development education has until now emphasized problematic situations in developing countries with a global perspective. Thus, the perspective of having learners connect the immediate community problems to the global issues and imagine a model for a new society has not been sufficiently recognized. Nor has it been actively put into practice, and consequently perhaps at the level of rationale, practice is going around in circles. Yamanishi (2008) appealed to the importance of developing mutually reinforcing approaches that move "from global to local" and "from local to global."

Development education has been searching for a long time for ways that learners can participate on a global scale. Its attention is now focused on the immediate local areas, with the notion that it is critical to nurture children's identity in the formation of multilayered regions within social participation learning. If this perspective is applied to the introduction of social participation learning in geography education, it may be necessary to define that identity as the fundamental core of geography education.

9.4.2 *Makeup of Children's Identity in the Formation of Multilayered Regions*

The abilities and skills that make up children's identity in the formation of multilayered regions contribute to societal change in the following ways:

- The foundation includes abilities and skills developed through direct experiences in the immediate local area. Participation in social activities that support the development of local society is grounded in values and rationales of the citizens.
- In a contemporary society where globalization is increasing, students need abilities and skills that allow them to be stakeholders who make decisions on issues across a range of spatial scales, from local to global.
- Students develop abilities and skills to understand issues from a broad perspective through investigating how issues across spatial scales are intertwined in the search for optimal solutions. Through this learning process, children should realize that they are citizens of the local, national, and global society, and that they need to be engaged in decision making and societal contribution.

Despite these educational efforts, is it possible for children to relate personally to issues at different spatial scales, thereby teaching them that they should participate as actors? Delanty (2006), who elucidated the plight of communities in the contemporary world, argues that communities are not systematic structures, shapes, or forms with symbolic meaning, but rather a community is constructed from a dialog-like process. An attribute of today's world is that communication is more important than anything else, and communities can take part in various means of communication. For example, the expansion of social interaction through social media and the Internet translates to building new social bonds at a global scale. As a result, "the citizens of today can participate not only within the social framework of the nation state, but also in the global society and the local society" (Delanty 2004, p. 254). Although children's current relationships are primarily founded on face-to-face interaction in their immediate local area, it has become possible through the various forms of communication for their identity to emerge in the face of formation of multilayered regions, at many different spatial scales.

9.5 A Geography Curriculum That Adopts Social Participation

9.5.1 Thematic Learning of Social Issues

Within the considerations described above, let us imagine what a geography curriculum that incorporates social participation would look like. Here, the author consider the framework of learning guidelines across Social Studies (third to sixth grade), the geographical fields in junior high school, and Geography A and Geography B in senior high school.

To weave local social participation learning into the curriculum, one must begin to leverage community problems that require solutions as teaching materials. However, the circumstances in which such problems manifest vary by region; there may be multiple issues, and the means for solving them may differ. Curriculum developers should recognize that the translation of real issues into curricular materials may only provide a simplified representation of reality. Also, the integration of social participation learning into geography should consider that geography attempts to make students aware of the contemporary world while shaping their world image.

Until now, organizing principles of the geography curriculum have been regional and systematic geography instruction. As a result, the recently revised National Curriculum Standards combine regional and systematic geography education at each stage of student development. If we assume that the study of community problems is indispensable for social participation learning, then it is essential to address current issues that require solutions. This is what is called thematic learning. This approach goes beyond the established regional divisions and type classifications of geographic phenomena, on which regional and systematic geography education are based. An exemplar is the "Geographic Inquiry into Global Issues (GIGI)" course, developed at the University of Colorado (Kusahara 2001). In Japan, the Research Group for a Consistent Curriculum in Elementary School, Junior High School, and Senior High School of the Geographic Education Society of Japan (2006) suggested an original Geography Alpha course in senior high school that deals with global issues. These thematic learning courses were designed to encourage students to explore social issues arising in contemporary society, as a way to learn the contents of systematic and regional geography.

All community problems are isolated and unique to a local area, but expanded to a larger scale, commonalities provide a rich perspective of such issues (e.g., the environment, poverty, economic inequality, and human rights) across regions. Thus, thinking about an issue can be understood at a deeper level when examining them at local and larger scales.

9.5.2 Four Spatial Scales for Fostering Children's Identity

The forms in which social issues manifest themselves vary with spatial scale. The author propose that four spatial scales be included in a curriculum: (1) local area; (2) Japan (the nation); (3) international (including cultural regions, economic regions, and others); and (4) global. The most appropriate scale best represents the issues; however, there is no guideline for its selection. Flexible establishment of an appropriate study space for the understanding and solution of social issues is referred to hereafter as the "multiple focus method" (Takeuchi and Ōno 2002).

A geography curriculum with the multiple focus method contributes to supporting children's identity in the formation of multilayered regions, a goal of social participation learning in the local area. Advances in thinking about social participation learning are based on events in the immediate local area, in which students search for possible solutions. In this context, students think comparatively about similar problems in other areas of Japan and, by recasting these within the nation, they acquire abilities and skills while shifting scales. Depending on the community topic selected, consideration from an international or global perspective may be useful. Students will deepen their abilities and skills as they form understanding of regions across spatial scales.[7]

Figure 9.1 presents a geography curriculum that incorporates social participation within the context of the National Curriculum Standards. The central core focuses on community

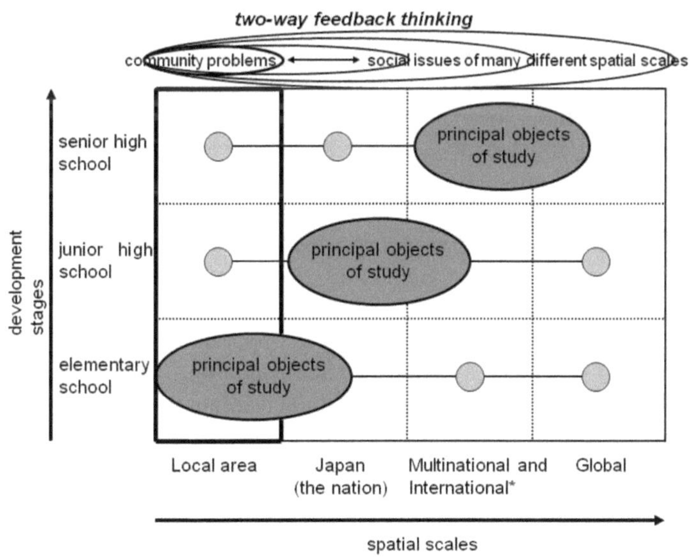

Fig. 9.1 Composition of a proposed Geography curriculum that incorporates social participation

[7] Research conducted with the same awareness goals presented in this paper is found in Hihara's theory of geography learning at multiple scales (Hihara 2005).

problems, which provide students with direct experience in their local area. To include social issues at all education levels, even though content is on national, multinational, and international scales, community problems must be actively adopted in teaching materials as illustrative examples of societal issues, or as objects for comparison.[8]

Not all geographic learning can be subsumed within social participation. For example, there are restrictions on the number of instruction hours, and opportunities are limited for children to fully engage in social participation learning.

9.6 Conclusion: Strategies for Instructional Design

To implement social participation learning in practice, improvements to the learning environment are indispensable (e.g., ensuring sufficient instructional hours, creating receptivity and networking in the local area, building a relationship of trust between the school and local community, and others). However, the reality of Japanese education makes success difficult to predict. Social participation learning is necessary for students to grow into self-sufficient citizens as they face societal problems. In conclusion, the author outline two important points from the perspective of instructional design for putting social participation learning into practice in the school system.

The first point is that children's awareness of society and development of their abilities as citizens are shaped by interaction with adults and people with various perspectives. Because of changes in society, such as the trend towards nuclear families, changes of industrial structure, and the loss of communal bonds, children grow into adults without awareness of a variety of values. As a result, children's views and thinking about society tend to be narrow and one-dimensional. Social participation learning provides children with a collaborative experience, in which they learn together and with adults from the local area. Those adults are citizens with varying occupations and values. An authentic learning experience may be critical for shaping students' awareness of society and nurturing their abilities as citizens (Hiroishi 2003).

The second point is to establish learning opportunities that are not fleeting involvements, but grounded in established and continuing committed partnerships. Current teaching practices of social participation learning involve one-time episodic events. However, community problems require solutions developed over time with committed dialog among members and related stakeholders. Thus, it is necessary to establish an exchange between people in the local community for extended periods, for example by linking to the course Period for Integrated Study.

[8]Because of space limitations, it is not possible to present details of the curriculum. The author hope to present specific proposals in another paper. Related to the proposals for the composition of a geography curriculum based on thematic learning, research findings of the Geography Education Creative Research Association have been made public (Ōno 2009).

References

Delanty G (2004) Gurōbaru jidai no shitizunshippu: atarashii shakai riron no chihei (Citizenship in a global age: society, culture, politics) (Trans: Satō Y). Nihon Keizai Hyōronsha, Tokyo
Delanty G (2006) Komyunityī (Community) (Transl: Yamanouchi Y, Ito S). NTT Shuppan, Tokyo
Fukushima T (1981) Chiiki no kadai to chiri kyōiku (Local region issues and geography education). Chirekisha, Tokyo
Hart R (2000) Kodomo no sankaku komyunitī zukuri to mijika na kankyō kea eno sankaku no tame no riron to jissen (Children's participation: the theory and practice of involving young citizens in community development and environmental care) (trans: IPA Japan Branch). Hōbunsha, Tokyo
Hihara T (2005) Atarashii chishi kyōiku (New regional geography education). Chiri kyōiku (Geogr Edu) 34:12–17
Hiroishi H (2003) Shimin kyōiku to shite no kyōdō keiken no kanosei (The potential of cooperative experience as citizen education). In: Ichimura N, Hayakawa M, Matsuura Y, Hiroishi H (eds) Keiken no imi sekai wo hiraku: kyōiku ni totte keiken toha nanika (Enlarging the world of meaning of experience). Toshindo, Tokyo, pp 55–77
Ida Y (2005) Shakaika kyōiku to chiiki: kiso, kihon no riron to jissen (Social studies education and local area: theory and practice of basic fundamentals). NSK Publishing, Tokyo
Ida Y (2006) Shakaika ni okeru chiri teki ginō no ikusei (Fostering geographic skills in Social Studies). In: Japan Social Studies Educational Association Publishing Project (ed) The challenge of social studies for opening a new era. Daiichi Gakushūsha, Tokyo, pp 112–121
Ikeno N (2001) Shakai keiseiryoku no ikusei: shimin kyōiku toshite no shakai ka (Fostering social formation abilities: social studies as citizenship education). Shakaika kyōiku kenkyū, bessatsu (J Social Stud Annu Rep) 2000:47–53
Imai T (1950) Nōson shakaika karikyuramu no jissen (Classes for social studies curriculum in rural regions). Maki Shoten, Tokyo
Itō R (1971) Shakaika kyōikugaku no kozo (The structure of social studies education). In: Utsumi I (ed) Shakaika kyōiku no riron to jissen: shakaika kyōikugaku genri (Theory and practice of social studies education: principles of social studies education). Aoi Shobō, Kyoto, pp 38–79
Kikuchi T (1960) Chiri gakushū no genri to hōhō (Principles and methods of geographic learning). Kaneko Shobō, Tokyo
Kikuchi T (1969) Chishi gakushū no kaizo to kihon teki sidou jikō (Reform of regional geography education and basic teaching points). Meiji Tosho, Tokyo
Kusahara K (2001) Gurōbaru mondai no chiri teki tankyū: GIGi no seikaku: 'shakai kogaku ka' toshiteno chirikyōiku (Geographic exploration of global problems– the nature of GIGI– geography education as 'social engineering'). Shakaika kyōiku kenkyū (J Social Stud) 85:11–23
Matsuno H (2004) Chiiki shakai keisei no shiso to ronri: sanka, kyōdo, jichi (Philosophy and logic of community social formation: participation, cooperation, and self-government). Minerva Shobō, Kyoto
MEXT (2008a) Chugakko gakushu shidō yōryō (The national curriculum standards for junior high school). Higashiyama Shobō, Kyoto
MEXT (2008b) Shogakko gakushu shido yoryo (The national curriculum standards for elementary school). Tokyo Shoseki, Tokyo
Ministry of Education, Science and Culture (1947) Gakushū shidō yōryō shakaika hen (I) (Shian) (The national curriculum standard, social studies edition I, (draft)). Tokyo Shoseki, Tokyo
Ministry of Education, Science and Culture (1958) Chugakko gakushu shido yoryo (The national curriculum standard for junior high school). Meijitosho shuppan, Tokyo
Ministry of Education, Science and Culture (1969) Shogakko shidōsho shakai hen (Elementary school teaching manual, social studies. Osaka Shoseki, Osaka
Ministry of Education, Science and Culture (1991) Kankyo kyōiku shidō shiryō (chugakko, koto gakko hen) (Teaching materials for environmental education: junior high school and senior high school. Ōkurasho Innsatsukyoku, Tokyo

Nihon chiri kyōiku gakkai shō-chu-kō ikkan karikyuramu kenkyū gurūpu (The Geography Education Society of Japan Study Group on Elementary School, Junior High School, and Senior High School Unified Curriculum) (2006) Chiri kyōiku ni okeru shō-chu-kō ikkan karikyuramu no kenkyū (Research on elementary school, junior high school, and senior high school unified curriculum in geography education). Chiri (Geogr) 51(4):68–77

Nishiwaki Y (2008) Shakai sanka noryoku to ikkan karikyuramu (Social participation abilities and the unified curriculum). In: Yamaguch Y, Nishiki T, Hatta F, Kobayashi M, Izumi T (eds) Chiri kyōiku karikyuramu no sōzō: shō-chu-kō ikkan karikyuramu (The creation of a geography education curriculum -a unified curriculum for elementary school, junior high school, and senior high school). Kokon shoin, Tokyo, pp 116–122

Ōno A (2009) Shō-chu-kō wo tsuranuku chiri karikyuramu no kōzō genri (Composition principles for a geography curriculum that extends through elementary school, junior high school, and senior high school). Chiri kyōiku kenkyūjo ronshū (Geogr Edu Res Insti Rep) 13:52–63

Satō K (2002) Kodomo ga sodatsu chiiki shakai (Community where children are raised). Tokyo Daigaku shuppankai, Tokyo

Shibusawa F (1998) 'Mijika na chiiki' wo ikashita nihon no shochiiki gakushū no jugyō kaizen (Improving classes in learning about the regional geography education of Japan that utilize the local area). Shakaika kyōiku kenkyū (J Social Stud) 80:9–20

Sudō Y (2006) Shokuryō jikyū ni tsuite kangaeru: nosagyō taiken, daizu saibai, chōsa katsudō wo tōshite (Thinking about food self-sufficiency– through farming experience, soybeans cultivation, and survey activities). In: Wada A, Asano T, Utsumi T, Ōno A, Sasagawa K, Fukuda Y (eds) Chiiki wo shirabe chiiki ni manabu: jizoku kanō na chiiki shakai wo mezashite (Researching regions and learning from them: towards a sustainable regional society). Kokon shoin, Tokyo, pp 71–80

Sumida M (2001) Chiiki shakai to kyōiku: kodomo no hattatsu to chiiki shakai (Local regional society and education: child development and community). Kyushu Daigaku shuppankai, Fukuoka

Takeuchi H (2006) Chiiki ishiki no keisei to shakaika: chiiki de manabu shikumi zukuri no hitsuyōsei (Forming local awareness and social studies—the necessity for creating mechanisms for studying the local area). In: Nihon Shakaika kyōiku gakkai shuppan purojekuto (ed) Shin jidai wo hiraku shakaika no chōsen (The challenge of social studies for opening a new era). Daiichi Gakushūsha, Tokyo, pp 98–109

Takeuchi H (2010) Shakai sanka gakushū nioite kodomo tachi ga kakutokushita mono: Shimin wākushoppu 'yomigaere! Shin-urayasu ekimae kōkyō hiroba daisakusen (2011nen)' ni sanka shita kodomo tachi eno tsuiseki chosa kara (What children gained in social participation learning from a follow-up survey of children who participated in the '2001 public square revival project for the front of Shin-Urayasu station"). Shin Chiri (New Geogr) 58(1):1–19

Takeuchi H, Ōno A (2002) 21 seiki chiri karikyuramu no sōzō (The creation of a geography curriculum for the 21st century). Keizai chirigaku nenpō (Ann Assoc Econ Geogr) 48(1):106–107

Takeuchi H, Sakuma A (1996) Kaihatsu kyōiku ni okeru 'sanka' no katei ni kansuru yosarsuteki kōsatsu: kōkō sei no jishu teki katsudō wo tegakari ni (Preliminary thoughts about the process of 'participation' in development education: based on senior high school students' self-directed activities). Kaihatsu kyōiku (Dev Edu) 33:85–99

Yamanishi Y (2008) Korekara no kaihatsu kyōiku to chiiki (Development education and local area from now on). In: Yamanishi Y, Kamijō N, Kondō M (eds) Chiiki kara egaku korekara no kaihatsu kyōiku (Development education from now on illustrated from local area). Shinhyōron, Tokyo, pp 4–16

Chapter 10
Geography Education as Education for Sustainable Development (ESD) in Japan

Shuichi Nakayama, Fumio Wada, and Jun'ichiro Takata

Abstract Education for Sustainable Development (ESD) in Japan has been taught in classrooms from elementary to senior high school since the Ministry of Education, Culture, Sports, Science and Technology (MEXT) introduced it as a compulsory study area in the revised National Curriculum Standards of 2008 and 2009. This chapter examines three driving forces of ESD and highlights exemplary practices that might be useful to teachers at every grade level. The first discussion presents characteristics and evolution of ESD in Japan and around the world. Second, international efforts to promote ESD are summarized from the Lucerne Declaration on Geographical Education for Sustainable Development (2007) by the International Geographical Union Commission on Geographical Education (IGU-CGE). Third, the theme and practice of the Associated Schools Project, initiated by UNESCO in 1956, are extracted to highlight the evolution of ESD. This reexamination of ESD formation to current instruction presents a roadmap for advancing geography education within ESD in Japan.

Keywords Education for Sustainable Development (ESD) • Lucerne Declaration • Sustainable society • UNESCO Associated Schools Project • United Nations Decade of Education for Sustainable Development

S. Nakayama (✉)
Hiroshima University, 1-12-14 Inokuchi-Myojin, Nishi-ku,
Hiroshima 733-0841, Japan
e-mail: snakayama@mist.ocn.ne.jp

F. Wada
Fukuyama University, Fukuyama, Japan

J. Takata
Gifu Shotoku Gakuen University, Gifu, Japan

Y. Ida et al. (eds.), *Geography Education in Japan*, International Perspectives in Geography 3, DOI 10.1007/978-4-431-54953-6_10

10.1 Introduction

Current issues among Japanese geography educators are threefold: (1) implementation of the revised National Curriculum Standards in 2008 and 2009 by the Ministry of Education, Culture, Sports, Science and Technology (MEXT), (2) a declaration that promotes Education for Sustainable Development (ESD) made by the International Geographical Union Commission on Geographical Education (IGU-CGE), and (3) UNESCO's development of a roadmap for geography education that promotes ESD in Japan.

Geography education is facing a new era, not only in Japan but internationally. The challenge that the new geography education faces is one of fostering student talents that can contribute to the building of a sustainable society within the new framework of ESD, under the United Nations Decade of ESD from 2005 to 2014.

The purpose of this paper is to first examine the direction of reform that sustainable development instruction demands from geography education, as well as the movement of IGU-CGE in response to the needs for education based on that development. Third we discuss the influence that UNESCO education, which has proposed the basic principles of education for sustainable development, has had on geography education. The discussion is aimed at clarifying optimal ways to reform ESD within geography education in Japan.

10.2 ESD: Driving Reform of Geography Education

10.2.1 Origins and Goals of ESD

The conceptual origins of ESD can be traced to the UN Conference on the Humanity and the Environment, held in Stockholm during 1972. However, the official beginning of ESD was at the UN World Summit for Sustainable Development (WSSD) in Johannesburg in 2002. That is where the Japanese government, together with NGOs, proposed that the period 2005 to 2014 be recognized as the United Nations Decade of Education for Sustainable Development (UNDESD). That same year, the proposal was approved unanimously by the UN General Assembly. This led to creation of the concept "building a sustainable society," which was introduced into the revised Japanese National Curriculum Standards. This was the beginning of a worldwide education reform movement proposed by Japan. When the UNDESD was approved by the UN General Assembly, UNESCO was designated as the lead agency. UNESCO drew up the International Implementation Scheme for the UN Decade of ESD, called "UNESCO IIS for UNDESD 2005," and obtained approval from the United Nations. This implementation scheme requested that each UN member state introduce ESD into their national curricula.

In Japan, the Liaison Council of Ministries and Agencies (hereafter referred to as the liaison council), which was concerned with the UN Decade of ESD, was established in the Cabinet Office in 2005. In 2006, the liaison council composed the "UN Decade of ESD Implementation Plan for Japan" (referred to hereafter as the Japan ESD implementation plan). In accepting this plan, the government referred to ESD as "education for sustainable development." The government designated it as one of the important priority measures for the next 5 years in the Basic Plan for Promotion of Education, published by MEXT (2008), based on provisions of Article 17 of the Basic Law on Education. Following this, the Education Ministry's Curriculum Council defined ESD as "learning that fosters the abilities to contribute to the building of a sustainable society," and added its implementation to the goals and content of related subjects in the revised National Curriculum Standards.

According to the liaison council in 2006, the goals of ESD were "to provide high quality education to all the world's people with a vision that the principles, values, and behavior needed for sustainable development are incorporated in all education and places of learning, and to bring about a change in behavior so that a sustainable future can be realized in the areas of the environment, the economy, and society" (Liaison Council of Ministries and Agencies 2006, p. 3).

10.2.2 ESD Introduced into Revised National Curriculum Standards

After some debate by the MEXT Curriculum Council, ESD was introduced into many related courses and subjects as learning to build a sustainable society. A number of subjects had parallel goals and content, including Social Studies and natural science in junior high school, and in senior high school, the courses Geography A and World History A and B. Science, Home Economics, Physical Education, Engineering, Science, Math, and others shared common learning outcomes as ESD goals.

Overlap between ESD and "geographic fields" as defined in the National Curriculum Standards at the junior high school level was found under the heading "Consideration that is centered on environmental problems and environmental protection." This standard focuses on "thinking about how it is important to deal with environmental problems and protection in the local area in order to build a sustainable society." The importance of learning ESD is described in Civics standards, and we argue that understanding about the building of a sustainable society has a place in Social Studies, in general. This includes related subjects such as Geography and History, which should be considered alongside junior high school Civics subjects.

In the revised National Curriculum Standards for senior high school, courses that include the idea of a sustainable society include the Geography and History subject area (World History A, World History B, Geography A), Civics (Contemporary Society, Politics and Economy), Science (all subjects), Health and Physical Education, Home Economics (Introduction to Home Economics, General Home Economics, Living Design, Consumer Life), and vocational training fields in Engineering (Introduction to Environmental Engineering), Science and Math, and others.

10.2.3 Abilities That ESD Attempts to Foster

The liaison council sought to impart the following abilities to students: (1) understanding of the background of problems and phenomena, and a system of thinking that emphasizes a comprehensive and multifaceted way of looking at the world; (2) development of critical thinking and an ability to consider alternatives; (3) analysis of data and information; (4) communication skills; and (5) values related to sustainable development that include respect for people, diversity, the environment, and equality of opportunity (Liaison Council of Ministries and Agencies 2006, pp. 7–8). Among these skills, the first four were clearly indicated in the former National Curriculum Standards as necessary "life skills," and they are practiced in classes. However, the fifth one is an important perspective that is necessary in ESD.

To implement ESD in the classroom, the most important objective is to cultivate students' thinking and behavior that will foster a sustainable environment, the economy, and society for the future. This teaches all that the norms of thinking and behavior are connected in the ability to build a sustainable society.

10.3 IGU-CGE and ESD

10.3.1 The Lucerne Declaration on Geography Education for Sustainable Development[1]

The leading organization for international promotion of geography education is the IGU-CGE. We refer here to the Lucerne Declaration on Geography Education for Sustainable Development (referred to hereafter as the Lucerne Declaration), published by the IGU-CGE in 2007 (Haubrich et al. 2007). The learned community uses the international trend in geography education as leverage to promote ESD, arguing that its principles and those of geography education are similar.

The essential policies of ESD were established in the UNESCO IIS for UNDESD, the leading committee of the UN Decade of ESD (UNESCO 2005). This document explains the aims for the UN Decade of ESD to succeed, whereas the Lucerne Declaration states in its preamble that the UN Decade of ESD is an opportunity to confirm CGE involvement in ESD. The declaration identifies the principles of geography education as being ESD, and explains sustainable development in the following ways: (1) It points toward the future; (2) it is a concept of harmony between humans and nature; (3) it is a concept of fairness and equity between generations, nations, cultures, and regions; (4) in addition to social, environmental, and economic problems, ESD extends to global responsibility and political participation.

[1] Translation of the declaration into Japanese was by Ohnishi (2008).

The Lucerne Declaration insists that this paradigm of sustainable development should be introduced into geography education at every level in every region of the world and, in the final "declaration," the CGE strongly recommends these principles be the foundation for geography education. The Lucerne Declaration is composed of three sections, which discuss: (1) the contribution of geography to ESD; (2) standards for developing geography education curricula toward the goal of sustainable development; and (3) the importance of information and communications technology (ICT) in ESD within geography curricula. Additionally, it is important to understand that the IGU-CGE in 1992 promulgated the International Charter on Geography Education,[2] and within this it was made clear that sustainable development is an important concept in geography education. This was a pioneering foresight.

10.3.2 Key Points of the Lucerne Declaration

10.3.2.1 Contribution of Geography to ESD

The Lucerne Declaration and UNESCO IIS for UNDESD shares the same vision for the future, which is "a world where everyone has the opportunity to benefit from education and learn the values, behaviour and lifestyles required for a sustainable future and for positive societal transformation" (UNESCO 2005, p. 6). It is clear that most of the "behavioral themes" from the UNESCO IIS are geographical. These are closely tied to the context of the environment, water resources, agricultural development, sustainable consumption, sustainable tourism, understanding of other cultures, cultural diversity, climate change, disaster prevention, biological diversity, and the market economy.

The close tie between ESD and geography is illustrated by geography's examination of the "Human-Earth" ecosystem (Haubrich et al. 2007) as a concept for sustainable development. The meaning is "from the viewpoint of humanity's survival, the earth is a household, and we should consume no more than we can obtain from it" (Haubrich et al. 2007, p. 2). Geography ties together the natural sciences and social sciences while studying the entirety of the Human-Earth ecosystem; it analyzes the amount of resources and living space that the Earth provides for humanity and how much impact society has on the Earth's ecosystem.

As a strategy to support sustainable development, the Lucerne Declaration proposed a new set of values. Sustainable development is a combination of ecological, economic, and social balance through novel living styles, new industries, and improvements in patterns of consumption. The report also states that to create a new value system for individuals, life-long learning that includes geography education is necessary.

---

[2] Translation of the charter into Japanese was by Nakayama (1993).

The research potential of geography in sustainable development is its offer of geographic knowledge (i.e., an understanding of the shared existence of nature and humanity), geographic skills, and cumulative values and attitudes aimed at solving various regional issues. For the sake of sustainable development, the cooperative engagement of geography with other academic disciplines is certainly required.

10.3.2.2 Standards for the Development of Geography Education Curriculum for ESD

The Lucerne Declaration advised that the development of the ESD curriculum should be led by each nation. Thus, each country will create their own national curriculum, related to regional themes, regions, and study methods, rather than the organization of a global standard curriculum.

During curriculum development, it is important to pay attention to the major issues of the contemporary world. This includes problems of humans and nature that exhibit spatial and sustainable behaviors. Also, topics such as global warming, energy depletion, misappropriation of non-renewable resources, population trends, and problems of worldwide inequality should be the focus. In particular, the Lucerne Declaration states that students should be encouraged to take up and consider conflicting and contradictory themes related to environmental, economic, and social sustainability.

As curriculum developers, attention should be given to four considerations, in which students: (1) follow an approach to functions, systems, predictions, behavior, structure, and processes that use a geographic way of looking at spatial organization; (2) choose learning content that can be models of structure and process, in relation to topics that can convey the essence of problems; (3) understand that importance of the problems depends on whether they are personal, public, political, specialized, or economic; (4) are aware that their interests and viewpoints vary from those of their peers, and select societal issues that are diverse, contrasting, and multidimensional.

The following three criteria outline regions useful to study as these: (1) are models of structures and processes; (2) are useful for fostering insight and penetration; and (3) provide a balance of variety and contrast in regard to location, type, and scale. In making the choice of study region, the following should be considered: knowledge, experience, interests, and concerns will vary with student age; the political, economic, and spatial position of the region and importance of its ecology and environment; a diversity of scale from the local, regional and national to the multinational and global.

Study and instruction methods for ESD are captured by the following four criteria: (1) Study should begin with simple example cases and gradually become more complex in content and method; (2) related facts should be arranged so that they build on each other; (3) a structural approach should be used to frame the concepts, processes, theories, and social structure that cause spatial changes; (4) exemplary case studies should be used to link common themes across regions. Students' varying interests, concerns, and learning at different grade levels should be considered.

10.3.2.3 Importance of Information and Communications Technology to ESD in Geography

The Lucerne Declaration recognizes the contribution of ICT to teaching and learning goals of geography education in ESD. Active use of ICT in ESD gives learners direct insight to people who are most affected by sustainability issues (e.g., environmental problems), as well as fostering students' conceptualization and deeper understanding of the attitudes of people in different cultures who are facing similar problems. ICT also supports a multifaceted perspective that is necessary for fostering sustainable attitudes and that promotes a high level of thinking, comprehensive and critical. Moreover, the declaration concludes that use of ICT expands research in geography education when incorporated within ESD. There is also potential for online cooperation that promotes students' learning about problems across a global scale and different cultures.

10.3.3 Significance of Lucerne Declaration

The Lucerne Declaration is significant in the geography education community because it clearly articulates the contribution of geography to ESD. The fact that the behavioral themes of the UNESCO IIS are mostly geographical indicates the importance that geography holds in ESD. To further buttress the value of geography, the latest geographic information and research results should be incorporated into the educational program and curriculum.

The concept of the Human-Earth ecosystem (Haubrich et al. 2007), advocated as a research method, is a framework based on geographic principles to examine global issues core to ESD. Such similarities and overlap make a case for introducing this concept into geography education as well. ESD is praised for its interdisciplinarity; however, even before the views of Jones et al. (2010), geography was essentially a comprehensive discipline characterized by cross-disciplinary research. This further supports geography education as a most suitable core subject to include in and promote ESD curriculum development.

The Lucerne Declaration envisions that each country will tailor a national curriculum, reflective of regional issues. This approach differs from the International Charter on Geographical Education, which aimed for international guidelines that would serve as a global standard, although it is in line with the UNESCO IIS for UNDESD (2005) policy. The Lucerne Declaration sets the ultimate goals for geography education to be included in ESD for educating citizens. As such, these citizens who develop sustainability values, are aware of the need to work toward a sustainable society, and have the skills to make this a reality. These goals are precisely what are necessary to inform people's decision-making and the codification of laws in each country (UNESCO 2005).

Geography education in ESD can be viewed as connected to social change. Education and learning can include actions that show students the need to always respect individuality.

The Lucerne Declaration clearly lays out the principles of geography education as ESD, and we agree that it is very important to direct these efforts in Japan.

10.4 Development of UNESCO Education in Japan and ESD

10.4.1 UNESCO Education and ESD

The Basic Plan for Promotion of Education (MEXT 2008) recommends that ESD courses be modeled through the UNESCO Associated Schools Project (ASP). The UNESCO schools began with ASP in 1956, 5 years after Japan joined UNESCO. It is possible to characterize the ASP, which was aimed at international understanding and international cooperation under UNESCO leadership, as the first phase of UNESCO education. While ESD was the second phase, aimed at building a sustainable society.

Similarities between the basic idea of ASP and ESD are found in their highly specialized coursework and methods involving a participatory approach. These modules are both very comprehensive and interdisciplinary in its learning contents. In addition to lectures, learning methods emphasized group activities, investigation, and information sharing through reports and presentations (Nakayama 2007). The parallels have prompted a comparison that identifies ASP of the postwar period as a milestone that served as a forerunner of ESD.

When UNESCO education was introduced into Japanese schools, domestic education experienced a decline in student scholastic abilities, which became a major social problem. Internationally, the first phase of UNESCO education was related largely to Japan's return to the international community of nations, whereas the second phase was realted to obligations and contributions to the international community. Table 10.1 (Takata 2011) summarizes education activities and efforts within Japan and UNESCO education.

10.4.2 First Phase of UNESCO Education and Content of Research Subjects

The authors reconsider the significance of achievements in UNESCO education through educational experiments of the Associated Schools in Japan. In contrast to previous evaluations held from the mid 1970s onwards, practice in the Associated Schools "rapidly declined" and placed the evaluations for UNESCO education in

Table 10.1 UNESCO education and trends of education in Japan and internationally

Period	Position of UNESCO Education in Japan	Developments in domestic education in Japan	Developments related to UNESCO education
1950s to 1970s	First phase of UNESCO Education	• Decline of scholastic abilities	Return to International Society
	• Associated Schools Program (ASP) aimed at international understanding and cooperation	• Systematization of Learning content (from unified Social Studies to differentiated Social Studies areas)	• As part of the citizen's UNESCO movement, Sendai UNESCO Support Association becomes first established in the world
			• ASP schools served experimental and research purposes
1970s to 1990s	• Stagnation of UNESCO Education	• Education for returnee children led by MOE	• International Education Recommendation by UNESCO, 1974
		• Development education, global education, multicultural education (education rationales and theories of practice introduced domestically)	• Politicization of UNESCO : U.S. withdrawal in 1984 (return in 2003); UK withdrawal in 1985 (return in 1997)
		• Peace education and anti-discrimination education (given by the teachers' union, etc.)	
From the 1990s onward	Second phase of UNESCO education	• Decline of scholastic abilities	Responsibility to International Society (contribution)
	• ESD aimed at building a sustainable society	• Increase of classroom hours (reduction of the Period for Integrated Studies and strengthening of education in various subjects)	• Development making UNESCO ASP schools, etc. into model schools

the context of development education or global education.[3] Here, the authors rethink the meaning of the Associated Schools' activities in relation to coursework regulated by the National Curriculum Standards, especially in relation to geography education.

[3] Sato (2007, p. 216) stated that "At the same time that implementation in the Associated Schools was ideological, because its scope was restricted, it did not sufficiently spread. Moreover, because it was preceded by a lofty rationale, there was a great deal of disparity between the implementations in different schools." Evaluations of the Associated Schools Program have not all been positive.

The Ninth UNESCO Plenary Session in Paris during 1956 clearly set out an experimental plan for the ASP. Based on this decision, education for international understanding and international cooperation was carried out in the worldwide expanding network of the Associated Schools participating in UNESCO's educational experiment. This occurred under the study of human rights, other countries (e.g., understanding other countries, ethnic groups, and regions), and the UN (e.g., study of world problems and the role of the United Nations in solving them). Additional modules were added to Understanding between Countries with Different Levels of Economic Development in 1962, and to Development and Environment in 1973. In 1984, these were distilled into four subjects: World Problems and the Role of the United Nations in Solving Them, Human Rights, Other Countries and Other Cultures, and Humans and the Environment (Nagai 1985). Development and Environment, which formed a common denominator with ESD, were already study subjects in the 1970s.

The educational experiment of the Associated Schools in Japan, under the leadership of the Japanese National Commission for UNESCO and Ministry of Education (MOE) at the time, was only put into effect in six designated schools. This effort had a major role as forerunner to the implementation of education for international understanding. The Hiroshima University Attached Senior High School was one of the designated schools. In particular, the implementation of education in subjects like "Study of Other Countries" overlapped with the content of geography education. At senior high schools that were Associated Schools, this was developed as "Geography Education for International Understanding" in Geography B classes. Also, in the UNESCO East–west Cultural Exchange Project developed in the 10-year plan beginning in 1957, the aim was to foster mutual cultural understanding between East and West, and interest in the Southeast Asian region was heightened.

However, in the 1970s, with international movements seeking development and independence for developing countries and UNESCO beset by issues facing all of humanity (e.g., population and food supply, resources, energy, and the global environment), there was a major change in direction from emphasis on international understanding to international cooperation focused on problem solving.

It is notable that with Japan's adoption of the "Recommendation Concerning Education for International Understanding, Cooperation and International Peace and Education Relating to Human Rights and Fundamental Freedoms" (abbreviated as the "International Education Recommendation"), there was a specification of major problems common at the global scale and an appeal to foster student interest in solving those problems.

10.4.3 Relationship Between Associated Schools Practice and Coursework

Between 1954 and 1969, the Hiroshima University Attached Senior High took part in 17th educational experiments (thereafter, this was carried out as educational implementation from 1970 through 1982) (Nagai 1985).

The subject "Study of the World's Nation Groups," which was implemented in the 1963–1964 school year as the Study of Other Countries, had content that corresponded to "Nations and Nation Groups" of Geography B. Under the condition that Study of the World's Nation Groups would "be an experiment positioned as much as possible within the regular curriculum centered around Social Studies Geography B," (Hiroshima University Attached Senior High School Research Section 1966, p. 3), classes for Geography B were developed. Senior high school Geography B is based on regional geography offered in junior high school, a "systematic learning" course. As Nations and Nation Groups appeared as a new important topic in the revised National Curriculum Standards (Hiroshima University Attached Senior High School Research Section 1966), the topic of nations was introduced in the Geography B curriculum at this school (Takata 2005).

In 1956, in addition to the six existing ASP schools, 13 schools were designated as research schools. At these institutions, studies were focused on topics that "spotlighted as its goal education for the sake of international understanding and international cooperation, centered on Social Studies and based on the National Curriculum Standards, while giving due consideration to the character of the area where the school is located and, moreover, based on the local and age characteristics of the students" (Nagai 1985, p. 71).

From the beginning, the educational experiment of the Associated Schools was closely tied to the National Curriculum Standards, borrowing from it the framework and course outlines. In particular, the policy was to stress its role within Social Studies. Afterwards, the National Curriculum Standards were the framework for the learning content in which the Associated Schools were involved.

In 1961, there were 28 UNESCO Associated Schools in Japan: 17 junior high schools, 9 senior high schools, and 2 universities. Most newly participating Associated Schools emphasized education in practice, and carried out coursework based on either or both the National Curriculum Standards or extracurricular activities (Nagai 1985).

The Associated Schools were a pioneering attempt to put international education into practice. Module content reflected both the study topics designated by UNESCO and coursework designated under the National Curriculum Standards. In particular, Social Studies played a major role in the educational experiments, because its contents overlapped with the study topics proposed by UNESCO.

10.4.4 Development Towards Geography in the Second Phase of UNESCO Education

In the revised National Curriculum Standards, learning content related to ESD is designated as coursework. The education for international understanding and international cooperation that the Associated Schools undertook in the immediate postwar era was conducted within subject coursework in accord with the National Curriculum Standards. Since there are many parallels between National Curriculum Standards with the UNESCO education, the latter being an important education program, can be included in schools via the national standards.

In the Study of the World's Nation Groups, results suggest a tendency toward unifying the differentiated learning contents within Social Studies. In ESD's development towards geography knowledge, in response to the question of how to unify that differentiated content, once again a reconsideration of geography education is called for.

ESD is extremely comprehensive and interdisciplinary, and it serves as a practical education for the future since it is based on critical thinking and problem solving-oriented thinking. Urabe et al. (2007) stated that as a discipline that builds the society of the future, it is necessary to revisit the discussion of geography education in Japan.

10.5 Conclusion

The Basic Plan for Promotion of Education (MEXT 2008) established the advancement of sustainable development education as one of the important policy measures for the subsequent 5 years. The revised National Curriculum Standards also incorporated it as new learning related to the building of a sustainable society, demanding a serious response from the geography education community.

ESD is a worldwide educational reform movement that began with the Japanese government's proposal to the UN in 2002 at the World Summit for Sustainable Development in Johannesburg. Acceptance of the challenge to develop exemplary modules that include global phenomena is a major responsibility for those who first made the proposal. The field of geography education responded to sustainable development by becoming part of the International Charter on Geographical Education in 1992. Enactment of this charter was in the same year as the United Nations Conference on Environment and Development, where there was international agreement on the concept of sustainable development. In 2007, the IGU-CGE adopted the Lucerne Declaration of ESD and proposed the importance of worldwide implementation. Japanese educational policy in the postwar period has consistently accepted the rationale of the educational reform movement urged by UNESCO, and has woven it into national education policy. In essence, the ESD proposal to the UN was founded on the fruits of UNESCO education, particularly through the Associated Schools Project in Japan.

The promotion of sustainable development instruction within geography education in Japan is necessary. Japanese leaders sufficiently understand the proposals of the IGU-CGE and the thrust of UNESCO education in its entirety, together with the Japanese government's goals. Nevertheless, their involvement is necessary so that the results of the country's past efforts can become an approved and accepted model for the world. In the midst of international change brought about by political, economic, and social globalization, a golden opportunity to create and globally promote high-quality geography education through ESD promotion in Japan can contribute to building a sustainable society. This reality lies right before our eyes.

References

Haubrich H, Reinfried S, Schleicher Y (2007) Lucerne declaration on geographical education for sustainable development. In: Reinfried S, Schleicher Y, Rempfler A (eds) Geographical views on education for sustainable development. Proceedings of the Lucerne-symposium. Geographiedidaktische Forschungen 42:243–250. http://www.igu-cge.org/Charters-pdf/LucerneDeclaration.pdf. Accessed 8 Feb 2011

Hiroshima University Attached Senior High School Research Section (ed) (1966) Yunesuko kyoudou gakkou keikaku•kyouiku jikken houkoku 'sekai no kokkagun no kenkyuu' (UNESCO ASP education experiment report on 'study of the world's nation groups'). Kenkyu kiyou (Hiroshima University Attached Junior High and Senior High Schools) 11:1–13

Jones P, David S, Sterling S (eds) (2010) Sustainability education: perspective and practice across higher education. Earthcan, London

Liaison Council of Ministries and Agencies (2006) Waga kuni ni okeru 'kokuren jizoku kanou na kaihatsu no tame no kyouiku no 10 nen' jisshi keikaku (Japan's action plan for UNDESD). Kokuren ESD no 10 nen kankei shouchou renraku kaigi, Tokyo

MEXT (2008) Kyouiku Shinkou Kihon Keikaku (Basic plan for the promotion of education). http://www.mext.go.jp/a_menu/keikaku/080701/002.pdf. Accessed 12 Feb 2011

Nagai J (1985) Kokusai rikai kyouiku ni kansuru kenkyu (Research on education for international understanding). Daiichi gakushuusha, Hiroshima

Nakayama S (1993) Chiri kyouiku kokusai kenshou (International charter on geographical education). Chiri kagaku (Geogr Sci) 48(2):46–61

Nakayama S (2007) Jizoku kanou na kaihatsu no tame no kyouiku– kaihatsu kyouiku to no setten wo motomete (Education for sustainable development– in search of a contact point with development education). Chiri kyouiku fouramu (Hiroshima Geogr Edu Allian) 8:71–84

Ohnishi K (2008) Jizoku kanou na kaihatsu no tame no chiri kyouiku ni kansuru rutsuerun sengen (The Lucerne declaration on geography education for sustainable development). Shin Chiri (New Geogr) 55(3&4):33–38

Sato G (2007) Kokusai rikai kyouiku no genjou to kadai– kyouiku jissen no aratana shiten wo motomete (Present and future issues in the education for international understanding– in search of a new perspective on educational practice). Kyouikugaku kenkyu 74(2):215–225

Takata J (2005) Yunesuko kyoudougakkou keikaku 'kokusai rikai to kokusai kyouryoku no tame no kyouiku'– 'sekai no kokkagun no kenkyuu' wo jirei toshite– (UNESCO ASP's 'education for international understanding and cooperation'– 'study of the world's nation groups' as an example). Chiri kyouiku fouramu (Hiroshima Geogr Edu Allian) 6:37–48

Takata J (2011) Nihon no yunesuko kyouiku ni okeru ESD no ichizuke (The positioning of ESD in Japan's UNESCO education). In: Nakayama S, Wada F, Yuasa S eds Jizoku kanou na shakai to chiri kyouiku jissen (A sustainable society and the practice of geography education). Kokon-shoin, Tokyo, pp 26–32

UNESCO (2005) United Nations decade of education for sustainable development (2005–2014): international implementation scheme. http://www.ESD-j.org/document/Final1%20IIS.pdf. Accessed 8 Feb 2011

Urabe M, Uemura H, Diwa JP (2007) ESD wa nihon no kyouiku sisutemu no noizu ka? (Is ESD noise within the Japanese education system?). Chiri kyouiku fouramu (Hiroshima Geogr Edu Allian) 8:57–70

Chapter 11
Institutional Issues in Disaster Prevention and Geography Education in Elementary and Secondary Education in Japan

Koji Ishizuka

Abstract This paper examines how disaster prevention is treated in elementary and secondary education, as well as in geography education. Disaster countermeasures should be taught in a comprehensive way, with other subjects, such as health and physical education, and natural science associated with disaster education. Countermeasure fundamentals should be instilled as solid scholastic skills within related subjects, as outlined in the National Curriculum Standards. The greatest contribution of geography education to disaster prevention is to encourage schoolchildren to explore their local area through field surveys. Teacher training courses can foster fieldwork in classes, but it is important to expand learning opportunities so that geography teachers can easily lead field activities.

Keywords Disaster prevention • Field survey • National Curriculum Standards • School safety

11.1 Introduction

The Great East Japan Earthquake and Tsunami of March 11, 2011, which shook the Japanese archipelago and caused an unprecedented amount of damage, marked a change in Japanese lifestyles. The event was a clear indication that disaster education is needed in schools. This idea has been presented to the government from experts in the field and the government is supporting efforts to improve disaster countermeasures in schools. However, school resources are stretched because they are confronted with many other problems such as violent behavior, bullying, and absenteeism. Furthermore, although schools are places for education, they also double as emergency evacuation shelters for the local community.

Disaster education in Japanese schools dates back more than half a century (Shiroshita and Kawata 2007). In 1947, the National Curriculum Standards for

K. Ishizuka (✉)
Tokiwa University, 1-430-1, Miwa, Mito, Ibaraki 310-8585, Japan
e-mail: ishizuka@tokiwa.ac.jp

Y. Ida et al. (eds.), *Geography Education in Japan*, International Perspectives in Geography 3, DOI 10.1007/978-4-431-54953-6_11

junior high school Social Studies, addressed the question, "What can we do to reduce as much as possible the damage from natural disasters?" as one of the units for second grade students. A learning goal was to help students understand that there are many dangers suffered from natural disasters in Japan. For example, Chapter 1 of the New Geography Course Series Volume 8 textbook entitled "Current Geography," published by Asakura Shoten Press in 1956, is about natural disasters. The structure of occurrences, impacts, and countermeasures related to climate disasters, earthquakes, and other disasters is covered. Additionally, the monthly journal *Chiri* has dedicated special editions to natural disasters for many years and, more recently, a special edition on disaster education. Likewise, content on disasters is included in the new National Curriculum Standards.

In response to the frequency of natural disasters in Japan, the government and schools have made various efforts to protect children and to promote disaster education. The purpose of this chapter is to identify elements that make up safety and disaster education, how geography education examines natural disasters, and what considerations are needed before recommendations from research are implemented.

11.2 The Positioning of Disaster Education in Japan

11.2.1 An Outline of Japan's Disaster Countermeasures

The Japanese archipelago is long and narrow, stretching approximately 3,000 km from north to south. The climate is not uniform but most of the country is in the temperate zone, possessing an abundance of nature and striking changes through the course of the four seasons. The archipelago lies on the boundaries of four tectonic plates, including the Eurasian and Pacific, and is along the Pacific Rim's mountainous Ring of Fire where earthquakes and volcanic activity are frequent.

Table 11.1 documents natural disasters in Japan between 1891 and 2011 where more than 1,000 people have died. During the 1950s, there was frequent damage from major typhoons and large-scale earthquakes, but the number of victims found dead or missing greatly declined with time. This is attributed to the advancement of national land conservation, improvements in weather forecasting, upgrades in the means of disseminating disaster information, and strengthening of the disaster management system. Despite such alerts, Japan is always under the threat of disasters such as major earthquakes, violent winds, and heavy rain (Cabinet Office 2002a).

The Disaster Countermeasures Basic Act was enacted in 1961 in the wake of the Ise Bay typhoon of 1959. Before passage of this act, legislation was passed to provide emergency measures in response to large-scale disasters (e.g., Disaster Relief Act of 1947), on an ad hoc basis. The Disaster Countermeasures Basic Act systematized all these disaster responses toward promoting a comprehensive and planned disaster prevention government framework. The act codified the organization, planning, prevention, emergency response, restoration and recovery, and fiscal and financial measures related to disasters. This same law stipulated cooperation between the

Table 11.1 Major natural disasters in Japan since 1891

Date	Disaster (Earthquake magnitude)	Number of dead and missing
1891.10	Nobi Earthquake (M7.9)	7,273
1896.06	Meiji Sanriku Earthquake Tsunami (M7.1)	21,959
1923.09	Great Kanto Earthquake (M7.9)	142,807
1927.03	Kita-tango Earthquake (M7.5)	2,925
1933.03	Showa Sanriku Earthquake Tsunami (M8.3)	3,064
1934.09	Typhoon Muroto	3,036
1943.09	Tottori Earthquake (M7.2)	1,083
1944.12	Tohnankai Earthquake (M7.9)	1,223
1945.01	Mikawa Earthquake (M6.8)	2,306
1945.09	Typhoon Makurazaki	3,756
1946.12	Nankai Earthquake (M8.0)	1,443
1947.09	Typhoon hits Kanto, etc.	1,930
1948.06	Fukui Earthquake (M7.1)	3,769
1953.06	Torrential Rains in Kyushu, etc.	1,013
1953.07	Torrential Rains in Wakayama	1,124
1954.09	Typhoon Toyamaru	1,761
1958.09	Typhoon Kanogawa	1,269
1959.09	Typhoon Ise-wan	5,098
1995.01	Great Hanshin-Awaji Earthquake (M7.3)	6,435
2011.03	Great East Japan Earthquake (M9.0)	18,880

Note: The number of dead recorded in the Great East Japan Earthquake is based on the White Paper on Disaster Management, Cabinet Office (2002b)
Source: Cabinet (2002a)

various branches of government and that each establish their own disaster management councils and headquarters to implement disaster countermeasures. Moreover, the legislation mandated that various public organizations develop management operation plans (e.g., the Central Disaster Management Council, public corporations, prefectural and municipal governments). At present, disaster countermeasures are being developed in response to more than 50 separate but related pieces of legislation (Cabinet Office 2002a).

The lessons of the 1995 Great Hanshin-Awaji earthquake taught the nation that it is extremely difficult to prevent disasters in advance (Cabinet 2002b). The White Paper on Disaster Management notes that combined with the efforts of public institutions, we are now encouraging self-assistance, based on the principle that each and every person should protect themselves by themselves. In this fashion, people prepare in their normal daily lives for the possibility of disaster, undertaking the durability of their homes to withstand shaking, fastening furniture so it does not fall over, preparing in advance the things they will need to take in an emergency, and participating in drills. Along with protection by the policies and assistance of public institutions, the report points out the importance of self-assistance, and cooperative 'community assistance' in which people in the local community aid each other.

Disaster response is in principle the responsibility of public institutions, but in cases where the disaster affects a wide area, there is a limit to the capabilities of public assistance. There is also a limit to what individuals can do through their own efforts. In such circumstances, what is needed is cooperative 'community assistance,' or disaster response at the local community level.

11.2.2 School Safety and Safety Education

In Japan, the modern school education system began with the School System Decree in 1872. To ensure the safety of schoolchildren, efforts were centered on health and hygiene regulation.

At present, schools are confronted with a diversity of problems, but the emergence of 'school safety' at center stage dates to the School Health Assurance Act that came into effect in April 1958. This law, aimed at protecting and improving the health of students and staff (elementary through university levels), developed a health supervision system in the schools. In December of the following year, the Japan School Safety Association Law was enacted, which mandated, "Together with planning to spread and strengthen school safety, necessary allowances are provided for the injuries, illnesses, disabilities, and deaths of elementary and secondary schoolchildren under the supervision of the various schools in the compulsory education system." Thus, both supervision and funding allowances were systematized.

Thereafter, both laws were revised in accord with demands of the times. The Japan School Safety Association Law was superseded in December 2002 by the National Agency for the Advancement of Sports and Health, Independent Administrative Agency Act. Then, the School Health Assurance Law was revised in June 2008 as the School Health and Safety Act, which stipulated "school health assurance" related to standards for school environmental hygiene, and that school safety be part of a comprehensive safety plan. This law remains in effect.

Concerning school safety, in March 1998 the Ministry of Education, Culture, Sports, Science and Technology (MEXT) distributed a pamphlet nationwide entitled The Development of Disaster Education for the Cultivation of Zest of Life as a reference document. In November 2001, the government distributed another national document, "School Safety Education for the Cultivation of 'Zest of Life'" as a reference to school safety for use in kindergarten through high school. According to these materials, the goals of safety education are to foster in students the qualities and abilities to: (1) Take appropriate conscious decisions and behavioral choices, based on accurate thinking and judgment; (2) predict dangers and improve dangerous environments independently; (3) be able to contribute to the safety of their schools, families, and local communities, all in accord with the schoolchildren's level of development. Safety education consists of 'safety instruction' within school subjects such as Science, Social Studies, and Health and Physical education; as well as 'safety guidance' in class activities, student association activities, and school observances (MEXT 2010).

11.2.3 Large-Scale Disasters and Disaster Education

In Japan, natural disasters are frequent. When large-scale disasters occurred, the government immediately formed a committee of experts to operationalize efforts aimed at strengthening disaster countermeasures in the schools. For example, the Central Disaster Management Council, held in July 1995 after the Great Hanshin-Awaji earthquake, established the Expert Survey Association for Promoting a Citizens' Movement to Reduce Disaster Damage, and published the 'Basic Principles for Promoting a Citizens' Movement to Reduce Disaster Damage' in April 1996. To foster school safety, in June 1995 MEXT established the Survey Research Collaborators' Conference on Strengthening the Disaster Management System of Schools, and issued an interim report on "Strengthening the Disaster Management System of Schools" in November that same year. A plan for school disaster prevention and a response manual for teaching staff were published, while guidelines were released on how to strengthen the school disaster management system in a follow-up report in September the following year. Additionally, in July 2011 MEXT established the Expert Conference on Disaster Education and Disaster Management in the Wake of the Great East Japan Earthquake, which carried out an interim assessment in September of that year. In July 2012, based on the 'Plan for the Promotion of School Safety' the ministry released results of the 'Research Survey on the Responses of Schools in the Great East Japan Earthquake'. A final report was published on ensuring instructional guidance time for disaster education, systematic allocation, organized training for teaching staff, maintaining a preparedness, collaboration with the local community, and a disaster prevention manual (Expert Conference on Disaster Education and Disaster Management in the Wake of the Great East Japan Earthquake 2012).

Coming out of these expert conferences were a variety of suggested measures for strengthening disaster education, administration, and organized activities in the schools. In these meetings, disaster education meant the cultivation of students' abilities to respond to disasters, and the aims stated in the "Strengthening the Disaster Management System of Schools Follow-up Report", released in September 1996. Those aims were integrated into various documents, for example, the "Disaster Prevention Learning Handbook" (February 2007) assembled by the Kuroshio Education Summit (2007) including four prefectures (Chiba, Shizuoka, Wakayama, and Kochi). This handbook states the goals of disaster education as follows: (1) students can make precise judgments and act to ensure their own safety, in accord with their development level; (2) to position disaster education as a subject that values life, and is useful during times of disaster to ensure the safety of other people, groups, and the local community; (3) using the local community as teaching material and holding disaster prevention drills in collaboration with that community, to help students understand the mechanisms by which disasters occur and their history, the local community's natural environment, and its disaster management system.

To achieve these goals, the basic questions are how to learn about and respond to disasters. In particular, in 2006, the Basic Law on Education was revised and a new stipulation was added to include 'mutual collaboration and cooperation between schools, families, and local area citizens.' Since 2008, MEXT has been considering the creation of a system in which the entire local area assists school education through the 'local area headquarters project for aiding schools.' However, there are many issues given the different school characteristics, such as who will lead the tasks.

11.3 Improving Instructional Implementation and Guidance in Disaster Prevention

11.3.1 How Disasters Are Dealt with in the National Curriculum

The national Curriculum Standards explains how each subject should deal with disasters (MEXT 2012). Following is a summary of those related to geographic education.

1. Elementary school/Social Studies: In the third and fourth grades, the prevention of accidents in the community should be addressed, along with examples of disasters selected from among fires, wind, water, and earthquakes. Examine how related institutions can cooperate with local residents to prevent disasters and accidents. By fifth grade, students should learn about "an outline of the nation's landforms and climate, the lives of people in characteristic regions seen from the viewpoint of natural conditions, and the work on forestry resources and natural disaster prevention for the sake of conserving the nation's land."
2. Junior high school/Social Studies: Along with helping students understand the characteristics of Japan's national land area surrounded by oceans, geography instruction should also consider the characteristics of domestic landforms and climate, natural disasters, and address prevention measures. Also, preventive measures in response to local natural disasters should be examined in the context of the natural environment and the regional characteristics of Japan in comparison to the world. At the same time, lessons in the survey of nearby surroundings should focus on phenomena and facts that are characteristic of the local area. These should be related organically to other phenomena, and the character of the local area should be examined. In relation to disasters, understanding can be deepened through field surveys.
3. Senior high school/Geography A and Geography B: Geography A is a subject that examines world issues as well as those in the local area. Within the standard geographical consideration of various local issues, students are expected to develop understanding of the relationship between the natural environment and natural disasters, as outlined in the newly established unit focusing on "the natural environment and disaster prevention". Together with instilling geographical skills that are connected to everyday life such as reading topographic and hazard maps, learning should be tailored to raising the awareness of disaster prevention. Geography B teaches geographical skills through the reading and construction

of many different types of maps and examination of various issues in the modern world, from systematic geography to regional geography approaches. Additionally, the use of maps and field surveys should reflect features of the school community.

The National Curriculum Standards include content about disasters that should be imparted through various means in classroom instruction of Social Studies and Geography. Besides these, there are explanations on how disasters should be dealt with in moral education, Health and Physical education, Science, Basic earth science, the Integrated Studies, and special activities. It is important that every class subject provide guidance on disasters.

11.3.2 Implementing Geography Classes on Disaster Prevention

The fundamentals on how geography treats disaster education do not differ from how other phenomena and facts are taught. The basis of the geographical way of seeing and thinking about the world includes the inquiry process with questions "Where is it? What is it? How can it be seen? Why is that? Is that true?" Field survey offers students indispensable opportunities to practice these geographical techniques. Simply working at a desk does not help students attain a geographic mindset. However, when students are carrying out a local survey, it is possible that awareness of their community increases, prompting them to take on local problems as their own.

The 1960 special edition entitled "The Sanriku Tsunami," published in the monthly journal *Chiri*, mentioned ties between geography and disasters. Since then, this journal has often featured disasters in special editions, recently in issues devoted to disaster education (August 2007), "geographic education standing up against earthquake disaster" (February 2009), "the Great East Japan earthquake" (June 2011), and "thinking about disasters and children" (May 2012). These issues have reported on topics such as the significance and status of geographic education and disaster prevention, as well as classroom implementation.

In the example below, I report on the classroom implementation experience of a teacher at Higashi Katsushika Prefectural High School in Chiba. Yuji Koseki undertakes instruction with enthusiasm and does progressive work. The following synopsis is based on an interview with him.

Instructional Goals At Koseki's former teaching post at Chiba Nishi High School, his goal in the Geography B class was to link the natural environment and a field survey. Through learning about regional characteristics of natural disasters, his aim was to instill in students the means of coping with such disasters.

Lesson Plan Elements (Spread Over Six Classroom Hours) (1) In the classroom, understanding regional characteristics of natural disasters; (2) Students survey their routes to school; (3) Learning about the local area using disaster prevention maps;

(4) Survey of damage in historically flooded areas; (5) Students conduct field surveys at a time of heavy rain and flooding; (6) Combining and collating survey reports, and identifying dangerous places; (7) Creating a flood hazard map; (8) Performing a safety assessment of the school.

Class Evaluation and Development The strong relationship between natural disasters and landforms and climate characteristics in the local area become a context for the project. In the project planning phase, students consider such questions as, "What kinds of natural disasters should we be worried about at Chiba Nishi High School?" By having students research their immediate surroundings, their desire for learning is increased. To raise students' awareness of dangers and to foster their ability to adapt to hazards, GIS and other means of three-dimensional expression are effectively used to simulate disasters. Koseki (2007) found that students' exploration of their local area and map construction based on data collection could be linked with disaster education.

Developing Activities Teachers at the school held a disaster prevention meeting that had 26 attendees, including educators from the school (the principal, vice principal, leading teaching staff, representatives from the students' association, and others), the community (parents' association, alumni association, local assembly, and others), and government (city office, local police, and others). As a result, Koseki is confident that they have realized several improvements, including the necessity for sharing awareness of disaster prevention, reexamination of the disaster prevention manual and evacuation drills, confirmation of actors' roles, identification of inadequate disaster prevention facilities, issues over the use of maps, and the importance of having all related actors and groups meet regularly to confer.

11.3.3 Classroom Realities and Ways to Improve Instruction

The National Institute for Educational Policy Research (2003) carried out a survey entitled "Research on the Curriculum," regarding the implementation status of the educational process, asking questions about how the goals and content of each subject specified in the National Curriculum Standards were being implemented in classrooms. The objective of the survey was to understand the actual learning situation to improve the educational process and means of future instruction.

In 2003, a written questionnaire was administered to both students and teachers about Social Studies and Geography learning in the fifth grade of elementary school, the first and second grades of junior high school, and the third grade of senior high school. Many r students in the lower grade answered that they liked Social Studies and Geography and believed them to be important. However, this trend declined with grade level increase (Table 11.2). If we look at types of classes, more than half of elementary school classes required schoolchildren to present information that they looked up at the library, and to incorporate observations and survey findings. In junior high school, the proportion of students taking geography declined by

Table 11.2 Questions to schoolchildren (percentage of "yes" responses)

School category	Grade level (year)	I like to study Social Studies and geography	Studying Social Studies and Geography are important
Elementary School	5th	54.7	84.6
Junior high school	1st	53.5	70.4
	2nd	53.5	66.9
Senior high school Geography A	3rd	32.0	58.0
Senior high school Geography B	3rd	43.6	63.6

Source: National Institute for Educational Policy Research (2003)

Table 11.3 Questions to teachers (percentage of "yes" responses)

School category	Grade level (year)	I teach classes involving library use	I teach classes involving experiences such as observations and surveys	I teach classes where students report on things they have looked up
Elementary School	5th	66.7	56.8	81.1
Junior high school	1st	28.6	19.2	45.6
	2nd	25.0	17.7	41.6
High school Geography A	3rd	8.3	10.8	8.6
High school Geography B	3rd	11.2	13.4	12.8

Source: National Institute for Educational Policy Research (2003)

half, and teachers in senior high school did not incorporate practices of geography at all (Table 11.3).

With the reality of classrooms as context, the following suggestions are given to improve instruction:

Elementary school, fifth year:

- Learning names and locations of islands and prefectures that make up the national land area
- Learning that the use of multiple resources can be related (e.g., statistical materials and distribution maps)

Junior high school, geography subject area:

- Learning locations and distributions of geographic phenomena, using maps and globes
- Learning geographic knowledge that forms the basis of geography learning

Senior high school Geography A:

- Learning that relates facts and phenomena to specific places
- Learning the use of maps and globes
- Learning to compare the local area to other regions and multiple types of data resources

Senior high school Geography B:

- Learning how to construct maps and graphs of geographical phenomena
- Learning that examines location and spatial extent of map elements
- Learning that relates different geographical phenomena and seeks cause-and-effect relationships

The survey findings about student and teacher reactions to learning and teaching can inform improvement of the fundamentals in geography education. The exceptional classroom implementation of Koseki was introduced above. Nevertheless, looking at the data, one may wonder if those responsible for teaching geography are knowledgeable about the ways of looking at and thinking about geographical phenomena or are adequately trained in geographical techniques. What is the situation of teachers on the "front line"?

The discussion below is based on the "Basic Survey of Scholastic Instruction" (Benesse Educational Research and Development Institute 2010), and explores the realities of learning and teaching at all school levels and teacher awareness.

The school condition in elementary schools is often marked by teachers who leave for work early and return home late. Teachers are at school an average of 11 h and 29 min/day. Only 33.8 % have both Saturday and Sunday completely off. Junior high school is similarly demanding, and teachers are at school an average of 12 h and 3 min/day, with only 8.2 % enjoying the weekend off. In the case of senior high schools, teachers leave for work at 7:43 a.m. and leave the school at 6:59 p.m., being at school for 11 h and 16 min every day. Only 7.6 % of these teachers say that they almost never go to work on Saturdays and Sundays. With increasing grade level, teachers spend more time at school, and work outside of normal hours increases.

Teachers in elementary, junior high, and senior high school all shared the feeling summarized by the statement, "There is a lot of school-business paperwork that has to be written up, and I can't find enough time to prepare teaching materials (Hyogo Prefectural Maiko High School 2012)." In senior high schools, teachers agree that "the educational administration has no grasp of the actual situation on the ground in the school". This is the number one complaint (Table 11.4). Teachers' work in the

Table 11.4 Teacher concerns, by school category

School category	Concerns (percentage in agreement)
Elementary School	Do not have enough time to prepare teaching materials (91.3 %)
	There is a lot of paperwork that must be done (84.2 %)
	The administration does not grasp conditions on the ground (76.9 %)
Junior high school	Do not have enough time to prepare teaching materials (81.3 %)
	Many students have not mastered the learning content of elementary school (80.9 %)
	There is a lot of paperwork that must be done (76.9 %)
Senior high school	The administration does not grasp conditions on the ground (78.7 %)
	There is a lot of paperwork that must be done (72.0 %)
	Do not have enough time to prepare teaching materials (65.1 %)

Note: Percentages that "strongly agree" or "agree somewhat"
Source: Benesse Education Research and Development Institute (2010)

schools does not end with teaching class, as they are involved in all kinds of activities like special education and students' clubs. As evidenced by the hours they work, their days off are almost nonexistent.

11.4 Issues in Disaster Prevention and Geography Education

11.4.1 Direction of Disaster Education

What form should disaster education take in schools? If school education is learning from textbooks, how can disaster prevention be systematized to cope with a wide range of disasters? A limitation of disaster prevention education is that its implementation in schools is very broad, from natural science to the humanities and Social Studies. What is needed is a comprehensive approach to disaster education. Furthermore, there are underlying problems that require solutions, such as how disaster education in the schools can be related to social education.

Society makes demands on school education. In response to the advent of an information-based society and the building of a sustainable society, there are many voices calling for information literacy and a responsibility to the environment. As a result, the course named Information, are now being introduced in senior high schools and the Integrated Studies has been established as a compulsory subject. Frontline educators are overwhelmed with these new learning movements, however, initiatives on disaster education are being taken. For example, Hyogo Prefectural Maiko Senior High School (2012) started the first department in Japan on environment and disaster prevention. Their website states, "There were no precedents anywhere in the country for a department specializing in disaster prevention, and in putting together the educational process, there were repeated trials and errors. The teachers responsible for leading this study repeatedly visited universities and attended disaster prevention conferences and seminars, read books and articles, and sought the advice of government disaster prevention sections, and amassing all this together somehow or other worked it up into the shape of a curriculum."

A variety of demands are being placed on school education. There is not enough time in the school day for teachers to devote to students. Moreover, the teachers are burdened with administrative tasks, summarized by "There is a lot of school-business paper work that has to be written up, and I can't find enough time to prepare teaching materials (Hyogo Prefectural Maiko High School 2012)." There is a real need for those involved in school administration to listen much more to the voices of teachers who are in the classroom.

The national curriculum explains how to include disaster-related topics within each subject, such that students are taught with sufficient rigor to develop solid disaster preparedness skills. Geography education, in which students learn about their communities and local areas, is capable of introducing disaster prevention in such a manner that problems become relevant to students. Geography provides a great way to contribute to disaster education.

11.4.2 Issues in Geography Education

Even if disaster education is taught across subjects, there are issues remaining. This section focuses on challenges within geography education. The Research on the Curriculum report (Sect. 11.3.3) indicated the need for improvement of basic instruction in geography education, similar to warning bells sounded by other experts. Niibori (2006) discussed the crisis in senior high school geography, pointing out the importance of making geography classes easy to understand so that teachers of other subjects can teach them. Similarly, Takizawa (2009) was concerned about the problem of training teachers and the vicious cycle in education. One of the issues in geography education is teacher training, Takizawa states this is especially the case when college students who have no actual experience in field surveys and other geographic activities that require geographic skills, when students who are not instilled with a geographical way of looking at and thinking about things, then go on to graduate and become teachers of Social Studies and Geography. In such cases it is difficult for them to teach geographic skills, and one can imagine that it is hard for them to communicate to their students the fascination of studying the local area, and hard for their classes to generate interest in and concern for the local area. Moreover, Ida et al. (2012) identified existing problems throughout elementary, junior high school and senior high school geography education, which are the needs to set out learning content that includes value judgments, deliberate decision-making, and social participation in the learning process, and to build up a continuity with the use of maps and other skills.

Thus, there are many issues in geography education, but to make disaster education effective, the question is, what can we do to help students think of disasters as their own problems? Along with teaching about the characteristics of disasters and the behaviors for avoiding them, as specified in the National Curriculum Standards, geography education is asked to provide classes in which students explore their own local area through activities such as field surveys of the immediate surroundings. It is necessary to train and support geography teachers who can stimulate students' interest in the subject. In order to do that, it is necessary to strengthen learning opportunities so that Social Studies and Geography teachers in the classroom can more easily carry out field surveys and other projects.

11.5 Conclusion

This chapter considers strategies to improve disaster prevention in elementary and secondary education, as well as issues in geography education. Japan is situated along the Pacific Rim's Ring of Fire, and the country is frequently threatened by natural disasters such as volcanic eruptions and earthquakes. The Great East Japan Earthquake and the Great Hanshin-Awaji Earthquake, dramatically demonstrated the importance of disaster education.

Assuring the safety of children in school is mandated by the School Health and Safety Act. School safety began with hygiene supervision, and was then extended to health education and extended further to safety education. Each time a large-scale disaster struck, the government made efforts to strengthen disaster countermeasures in the schools. However, schools are confronted with a wide range of problems. Because disaster countermeasures must be tackled comprehensively, it is essential for not only the schools but also for families and the community to collaborate. For effective disaster education, class content should be aligned with the instructional content specified in the National Curriculum Standards. As a result, the knowledge and skills acquired in each subject can be used when students are faced with a disaster.

The greatest contribution that geography can make to disaster education is to provide learning opportunities where students explore their own living environment with such tools as surveys. However, teachers face many other problems such as chronic absenteeism and, as substantial overtime work increases, they have no time to prepare teaching materials. However, the National Institute for Educational Policy Research showed that there are signs of improvement in the core content of geography education, based on results of the Research on the Curriculum. To improve these circumstances, the design of teacher training courses is also relevant, but it is also important to expand in-service opportunities so that teachers involved in geography education can easily conduct activities such as field surveys.

Disaster education should be promoted. One hopes that school administrations promote this despite the realities of the education system, to maintain a firm grasp of what is most important for students and frontline educators.

References

Cabinet Office (2002a) Wagakuni no saigai taisaku (Disaster management in Japan). http://www.bousai.go.jp/panf/saigaipanf.pdf. Accessed 23 Aug 2012

Cabinet Office (2002b) White paper on disaster management. http://www.bousai.go.jp/kaigirep/hakusho/h14/index.htm. Accessed 23 Aug 2012

Benesse Educational Research & Development Institute (2010) Fifth basic survey of scholastic instruction. http://berd.benesse.jp/berd/center/open/report/shidou_kihon5/sc_dai/index.html. Accessed 23 Aug 2012

Expert Conference on Disaster Education and Disaster Management in the Wake of the Great East Japan Earthquake (2012) The final report of an expert conference on disaster education and disaster management in the wake of the Great East Japan earthquake. http://www.mext.go.jp/b_menu/shingi/chousa/sports/012/toushin/1324017.htm. Accessed 23 Aug 2012

Hyogo Prefectural Maiko High School (2012) About the environment and disaster mitigation course. http://www.hyogo-c.ed.jp/~maiko-hs/bosai/kanbou1.htm. Accessed 23 Aug 2012

Ida Y, Yoshida K, Hirasawa K, Asakawa T (2012) Nihon no gakko chiri kyoiku ni okeru genjo to kadai (The circumstances and problems in geography education, Japan). E J Geogr 7(1):3–10

Koseki Y (2007) Anzen kyoiku ni okeru GIS no yuyosei: bousai, bouhan shido ni koken suru chiri kyoiku no arikata (An effective use of GIS in safety education–an instructive contribution of geography to disaster and crime prevention). Chiri Shiso 48(2):16–22

Kuroshio Education Summit (2007) Disaster prevention learning handbook. http://www.pref.shizuoka.jp/kyouiku/kurosio/bousai3.html. Accessed 23 Aug 2012

MEXT (2010) 'Ikiru chikara' wo hagukumu gakko deno anzen kyoiku (School safety education for the cultivation of zest of life). http://www.mext.go.jp/a_menu/kenko/anzen/1289310.htm. Accessed 23 Aug 2012

MEXT (2012) Shin gakushu shido yoryo (The new national curriculum). http://www.mext.go.jp//a_menu/shotou/new-cs/youryou/index.htm. Accessed 23 Aug 2012

National Institute for Educational Policy Research (2003) Research on the curriculum. http://www.nier.go.jp/kaihatsu/kyouikukatei.html. Accessed 23 Aug 2012

Niibori T (2006) Chiri ga nakunaru? Chiri kyoshi ga inakunaru? (Will geography courses disappear? Will geography teachers disappear?). Chiri (Geogr) 51(6):22–27

Shiroshita H, Kawata Y (2007) Gakushu shido yoryo no hensen katei ni miru bousai kyoiku tenkai no kadai (Institutional problems in disaster education in compulsory schooling analyzed from the historical transition of the course of study). Shizen saigai kagaku (J Jpn Soc Nat Disaster Sci) 26(2):163–176

Takizawa Y (2009) Chiri kyoiku no genjo to kadai (The present circumstances and problems in geography education). Ochanomizu chiri (Ann Ochanomizu Geogr Soc) 49:2–9

Part III
Challenges to Teacher Training and Geography Education

Chapter 12
University Geography Departments and the Teaching License in Japanese Geography Education

Koji Ohnishi

Abstract This paper provides a historical background for the current state of geography education in Japan, focusing primarily on the role of Japanese universities in the teacher training process. There are several problems in the teacher licensure process in Japan. For elementary school teacher training, candidates are not required to have taken a geography class. Prior field research and map use training is not required for teachers licensed to teach secondary school Social Studies and the Geography and History subject areas. Many of these university programs fail to provide sufficient geography education to pre-service teachers. As a result, newly licensed educators are unable to competently teach the subject.

Keywords Geography class • Geography department • Teaching license • University

12.1 Geography Departments in Universities

The geography departments in Japanese universities are usually within one of three faculties: Literature, Science, or Education. Recently, the department of geography may also be within other faculties, such as the faculty of Environment. In the past, there were regional differences in the location of university geography departments. In 1907, geography was within the Faculty of Literature at Kyoto University, whereas it was within the Faculty of Science at The University of Tokyo. Following this trend, national universities in western Japan normally included geography departments within the faculty of Literature, whereas the faculty of Science housed them in the eastern part of the country (Ukita et al. 2001). Currently, geography departments are not set up according to regional jurisdictions.

It is important to have geography classes at universities that students obtain teaching licenses for elementary school education, Social Studies education in junior high school, and Geography and History education in senior high school.

K. Ohnishi (✉)
University of Toyama, 3190 Gofuku, Toyama 930-8555, Japan
e-mail: ohnishik@hmt.u-toyama.ac.jp

Y. Ida et al. (eds.), *Geography Education in Japan*, International Perspectives in Geography 3, DOI 10.1007/978-4-431-54953-6_12

There are geography departments in all faculties of education and universities of education, where training of teaching candidates is provided.

Within private universities, geography departments were established before World War II as part of the junior high school teacher training course (Ukita et al. 2001). The number of students in these geography departments is larger than that of national universities. Students who study geography in these private universities raise the quality of geography education in Japanese society.

The Association of Japanese Geographers lists 246 universities with geography teachers and researchers. There are 778 universities in Japan, although only 31.6 % have geography faculties. Therefore, geography researchers and educators are a minority in Japanese universities.

12.2 The University of Education and Faculty of Education: Teachers' Training University

Japan has 47 prefectures, each with a University of Education or Faculty of Education that is part of the teachers' training program. To teach elementary school in the country, teacher candidates study at one of these institutions in order to qualify for a teaching license. Teachers of junior and senior high schools take courses in their subject of instruction (e.g., geography) and in education-related topics offered by the department of education and other relevant departments. Each university must provide specialized geography training to support pre-service teachers develop their knowledge and understanding so they can impart to their students the manners in which geography applies to our current society and its future development.

12.3 System of Teaching License

Teaching licenses are classified in accord with subjects and schools. The following detailed descriptions refer to university geography classes that fulfill licensure requirements.

12.3.1 Elementary School Teaching License

Elementary school teachers are responsible for instructing all main subjects, including geography; however, these teachers are not required to have taken geography classes in their university education. Approximately 21 % of pre-service teachers in the department of education had not taken a geography class, but were still successful in obtaining their elementary school teaching licenses (Shirai 1998). Some candidates who become elementary school teachers have few opportunities to learn geography skills and knowledge while at university. The problem is that some teachers have poor map skills and spatial thinking skills, which are important to teach geography in Social Studies at the elementary school level.

12.3.2 Junior High School Teaching License

License types are defined by subject area. Geography is included within Social Studies at the junior high school level. The teaching license at this level requires only two classes in geography at the university level, geography and regional geography. Both courses, optional for university students but mandatory for teacher candidates, are given in a didactic manner, which is not a practical way to impart map use or the field survey experience. Only 10 % of Social Studies teachers major in geography at university (Shirai 1998). As a result, there are few teachers in Social Studies who have rich set of geography skills within the junior high school education system.

12.3.3 Senior High School Teaching License

The teacher license for geography instruction, within the Geography and History subject area in high school, permits educators to teach World history, Japanese history, and Geography. World history is a compulsory subject in senior high school, whereas Geography and Japanese History are elective subjects. Therefore, some high schools do not offer geography classes at all. The Association of Japanese Geographers Commission on Geographic Education (AJG-CGE) reported that 26 % of high schools did not have any geography classes, and thus there was no opportunity for students to learn the subject (Usui 2008).

The Geography and History teaching license requires three courses total, one course each in human geography, physical geography, and regional geography at the university level. However, practical training classes for map use and field research skills are not required. The number of Geography and History teachers with a major in history is larger than those with a major in geography. Thus, teachers have many difficulties when attempting to effectively teach geography in high school.

12.4 Difficulties in Managing Senior High School Geography Course

It is difficult for teachers who have not majored in university level geography to teach it in senior high school. Musha (2000) made a unique investigation related to this topic. He analyzed how teachers prepared their lesson content, which aspects of geography they had difficulties with, and how they resolved the obstacles.

The findings indicate that teachers confessed difficulties in teaching such topics as climate, topographic maps, map projection, and field research. Teachers have few opportunities to study the mechanisms of climatology and require more technical instruction in this area.

Regarding topographic maps, teachers had insufficient instruction and experience to effectively read and interpret the maps. A map projection has mathematical aspects that require more in-depth formal instruction for educators to have confidence teaching the subject. Finally, there is a need for more field research training.

Textbooks in schools are authorized by the government. Their content covers various topics in geography, although the depth of coverage is limited by the fixed number of pages mandated by the government. Thus, the textbooks only provide a shallow introduction to the material. There is a need for textbooks or guidebooks tailored to non-geography majors in the teaching profession.

12.5 Geography in Universities and K-12 Geography Education

Teacher candidates can receive a license to teach geography having only taken a few geography classes at university. Once they become qualified teachers, they may have difficulty imparting geographic knowledge and skills to their students. Without a formal education background in geography, teachers lack understanding of regional phenomena and implications at local and global levels. As a result, their students receive an incomplete understanding of geography and are unable to effectively communicate Japan's role in world affairs.

Teacher training programs should create content for geography classes that meet the needs of teacher candidates as they prepare for the classroom. Universities have a responsibility to better educate teacher candidates by updating teaching materials and pedagogical strategies. For example, pre-service teachers should be required to learn basic geography skills like map reading and field research. The important role of geography in our global society cannot be overlooked. This positions educators as important vehicles for conveying the importance of geography in education, and for administrators to make effect changes in the education system.

References

Musha K (2000) Chiri wo senmon to shinai kyoin no koko chiri kyoiku ni taisuru ishiki: Niigata kennai no chosa wo motoni (Teacher's view of high school geographical education). Shin Chiri (New Geogr) 48(2):12–23

Shirai T (1998) Kyoin yosei ni okeru chiri kyoiku no genjo to kadai: daigaku kaikaku no naka no chiri kyoiku (Present status of and subjects in geographical education during teacher Training: Geographical education under the university reforms). Chirigaku Hyoron (Geogr Rev Jpn) 71A(2):104–112

Ukita T, Ikeda H, Todokoro T, Noma H, Fujii T (2001) Jio paru 21: chirigaku benricho (Geo Pal 21: guide for geography education). Kaiseisha, Otsu

Usui T (2008) Chiri rekishi ka kyoin no jittai to chiri teki chishiki teika no mondai ten (Geography and history teachers status in high school and students' geography knowledge decreasing). Gakujutsu no Douko (Trend Sci) 23(10):13–19

Chapter 13
The Association of Japanese Geographers' Social Action Program and Geography Education

Takahisa Izumi and Hiromi Iwamoto

Abstract The Association of Japanese Geographers (AJG) sponsors and supports activities for members as well as non-members. In this article, activities for non-members are referred to as the "social action program". It comprises two categories: (1) activities for the general public and (2) geography education for school teachers, and students. The former, which typically takes the form of public symposia, are designed to promote the value of geography in society and the public arena, whereas the latter category includes public fora for teachers, Geography Olympiad for high school students, and Children's Environmental Map Contest for elementary and junior high school students. Together, these outreach efforts aim to revitalize geography education in Japan. The AJG Commission on Geographical Education (AJG-CGE), established in 1998, plans many of these social action programs in collaboration with other academic societies and research groups.

Keywords Children's Environmental Map Contest • Commission on Geographical Education • Teacher education • The Geography Olympiad

---

Note: This paper was divided among the authors as follows: Section 1: Iwamoto; Section 2: Iwamoto; Section 3: Izumi; Section 4.1: Izumi; 2: Iwamoto; Section 5: Izumi; Abstract: Iwamoto. In Sections 1 and 2, Izumi added parts based on his experience as a member of the AJG-CGE.

T. Izumi (✉)
Senshu University Matsudo High School, 2-3621 Kamihongo, Matsudo, Chiba 271-8585, Japan
e-mail: izumi.ta@isc.senshu-u.ac.jp

H. Iwamoto
Nara University of Education, Nara, Japan

Y. Ida et al. (eds.), *Geography Education in Japan*, International Perspectives in Geography 3, DOI 10.1007/978-4-431-54953-6_13

13.1 The Context in Which the Association of Japanese Geographer's Social Action Program Is Needed

This chapter begins by explaining efforts undertaken by the Association of Japanese Geographers (AJG) to engage non-members, from the 1980s to the present. This is followed by a focused discussion of educational activities provided for school teachers, children, and students. Recent examples include the "Science and Geography Olympiad Japan Championship". This is a competition for high school students, jointly sponsored by the AJG, Japan International Geography Olympiad Committee, Japan Science and Technology Agency (JST), and other organizations. In another partnership, the AJG is working with the Society for Environmental Map Education (SEME) and other organizations to host the "Exhibition of Environment Maps of Our Immediate Surroundings," a project for all students. In-service training for teachers is provided by the AJG-CGE. A description of project implementation, results, and issues will provide further information on these projects.

The AJG directs and organizes social action program activities for non-members. In recent years, it has become customary to plan free public events for non-members. For example, at the general conferences, the association planned activities with a strong awareness of making a social contribution. This is not only true of the AJG, but also applies to the Human Geographical Society of Japan (HGSJ) and other geographic associations. Efforts to include the public in the social action program have been driven by two major issues. The first is that geography has become increasingly valued in society. Students who complete an undergraduate and graduate geography program in college are employed in diverse fields throughout society. Geography graduates demonstrate the usefulness of geography within their workplace and profession. Until recently, there has been little recognition, in an official or national way, for the skills and qualifications directly related to the study of geography. In essence, the application and value of geography is neither well known nor respected. To improve understanding of geography's contribution to society, the geographic associations should develop strategies to institutionalize qualifications related to geography to inform the general public.

The second issue is that geography enrollments in high schools are dwindling; a revitalization of geography education is badly needed. Enrollment levels began to drop in 1978 when the National Curriculum Standards for senior high schools chose Contemporary Society to be a compulsory subject and Geography was relegated to an elective course. In the 1989 revision of the National Curriculum Standards for senior high schools, the Social Studies curriculum was divided; a Geography and History curriculum was newly established, and World History A and B were made compulsory subjects. This had a major impact, causing high school enrollment levels in Geography to drop even further than before. This trend has continued to the present (Kohno 1996).

The AJG is active in lobbying the Ministry of Education, Culture, Sports, Science, and Technology (MEXT) to support geography education. In the past, the AJG made frequent requests that the previous Ministry of Education address the decline in geography enrollment. Efforts to make educational content more attractive are needed to revitalize geography education. Thus, support for geography teachers and direct impacts on students in the education system are equally important to improve the image of the geography field.

13.2 The Beginning of the AJG's Social Action Program and Its Course Up To the Present

13.2.1 Public Events

We describe here the background of how AJG's social action program began and its evolution to the present. Information was gathered from records of the board of directors and news articles published in the journal *Geographical Review of Japan.*

At a board of directors meeting during the autumn AJG general conference in November 1989, board members deliberated on holding an open public lecture at the 1990 general conference. The holding of open public lectures was approved and added to the association's bylaws. Since then, AJG's social action programs have often been held at the general conference. Table 13.1 summarizes AJG's public events during the 1990s.

The series of events in the 1990s laid the groundwork for AJG's involvement and plans to engage the public through its social action program. The social action program is the impetus to the implementation of geography education open public lecture series that began in 2001.

Thereafter, the open public lectures were presented during the general conferences in 2004 and 2005. Lectures were replaced with symposia from 2006 to 2012. Support for public engagement was seen at the autumn general conference at Nagoya University in 2010; as many as ten different symposia were advertised as "open to the public". Today, open public events are firmly established at AJG general conferences. However, the number of non-members who have participated in these symposia is unclear.

In addition to efforts by the AJG, other academic associations have provided free lectures to the public. For example, in November 1998, the Human Geographical Society of Japan HGSJ held a 50th anniversary commemorative celebration, and an open public lecture series was held at the Kyoto Conference Hall as part of the events.

Table 13.1 AJG's public events during the 1990s

Date	Theme	Location	Lecturer
September 1990	Geography Education: Looking Back and Looking Ahead—Based on practical examples	Joetsu University of Education's attached junior high school	Asakura Ryutaro
October 1991	The Problems of the Aging of Society and Depopulation, and Permanent Residency Measures: From the Standpoint of Geography	General Welfare Center, Matsue City	Fujita Yoshihisa, Shinohara Shigenori, and others
March 1997	Open Public Invitational Lecture	Tokyo Metropolitan University	

The general theme was "Internationalizing Social Space and Local Livelihood," and featured speakers Narita Kozo, Takeuchi Keiichi, and Sasaki Komei. This event initiated substantial consideration for including open public events such as seminars. In December 2001, in an event separate from the society's general conference, HGSJ held their first open public seminar, using the Osaka Prefectural Labor Center in Osaka's Chuo Ward. The theme was "Frontiers of Human Geography". Records of meeting announcements, printed in the *Japanese Journal of Human Geography*, show that the HGSJ has since held annual open public seminars separately from their general conferences, rotating venues through each of the prefectures of the Kinki District.

The series of open public events put on by the AJG and HGSJ have broadly targeted citizens. A preliminary conclusion is that the implementation of these events has encouraged investment in social action programs within the field of geography education.

13.2.2 Undertakings of the AJG-CGE

Since the AJG-CGE was established in 1998, it has actively implemented public events for school teachers and citizens. The AJG-CGE's goal is to enlighten people about the value of geography and to revitalize geography education.

One of the pillars of the AJG-CGE's social action program is the open public lecture series, held concurrently with the association's general conferences. The spring general conference in 2013 marked the 23rd public lecture series, which has been held without interruption since the autumn general conference in 2001.

At the 2001 autumn conference, the public lecture theme was "A Professional Talk about His Fieldwork—Knowing about the world through geography, No. 1: Turkey". These presentations were designed to be engaging for non-geographers and a general audience. Additionally, it was the intention to include an international aspect as a way to widely publicize the latest research findings in geography education. Lectures have been recorded for the following countries or regions, in order of presentation: Turkey, the United States, China, New Zealand, India, and South Korea. Not all themes were geographically focused. In the autumn conference in 2008, the theme was "Education for Sustainable Development and Geography Education". Since then, themes have had a regional or non-geographic focus, but the goal of the open public lectures remains committed to the revitalization of geography education (Shimura 2007).

13.3 Examples of Outstanding Recent Social Action Programs: Events for Children and Students

13.3.1 Geography Olympiad Japanese Championship

The Geography Olympiad Japanese Championship (referred to below as 'the championship') is divided into a preliminary selection stage and a final selection stage. The preliminary competition is held every year in mid-January, and the second stage is held annually in mid-March. As of 2013, the championship has been held seven times, and it has become firmly established as a geography education event every spring (Izumi 2013).

The Geography Olympiad Championship serves as a selection process for Japanese contestants to enter the International Geography Olympiad that is sponsored by the International Geographical Union's Task Force. Its purpose is intended to promote geography education in Japan by attracting high school students to compete in multiple geography skills and abilities, including areas of knowledge and concepts, thinking and judgment, and skills and expression. Simultaneously, students demonstrate their geographical knowledge while competing at an international scale, thereby highlighting the social value of geography. This is in contrast to the rote learning in Japan.

When Japan competed in its first international competition in 2007, there were only twelve Japanese participants. Thereafter, the number gradually rose to 867 in 2013, and those who actually took the exams reached 782. This increase may be attributed to numerous factors, including efforts at advertising the competition by the executive committee. With the cooperation of other sponsors and support organizations, the competition was designated by the JST as a Science Olympiad, similar to those for biology or earth sciences. It was also supported by MEXT, which together with JST support has dramatically increased the level of awareness about the Geography Olympiad. This has led people involved in geography and geography education from across the country to become members of the executive committee. They help set up exam centers and request local teachers to publicize the competition, as well as encourage their own students to take the exams.

Exam takers come from across Japan, from Hokkaido to Okinawa. The national attention given to the Geography Olympiad has contributed greatly to strengthening ties between geography education supporters across the country.

The exams used in the championship selection process consist of two parts, a multimedia test in the first round and written response test in the second round. Table 13.2 summarizes characteristics of the two parts.

After the championship competition, both parts are scored, data are aggregated, and a total score calculated. Prize winners are decided from the top scores for gold, silver, and bronze medals. From the gold medal winners, those with the highest total scores and within the top four places in the English part of the exam are chosen to represent Japan at the International Geography Olympiad. The awards ceremony is in late March at the spring AJG general conference, with many association members in attendance. Contestants are officially recognized as the representatives for Japan after award certificates and supplementary prizes have been presented by the director of the association.

The international competition comprises of the two exams mentioned earlier and a fieldwork exam. In the latter, teams of four students with mixed nationalities perform a field survey, then compile the results and present them orally. To prepare Japanese contestants for the fieldwork portion, geography education leaders gather the student representatives in mid-June for a 2-day workshop. There they receive practical guidance in field survey methods that include field observation and interviewing, collecting materials, and methods of analyzing results.

Table 13.2 Characteristics of the two exam parts

	First part	Second part
Type	Multimedia test	Written response test
Form	Fifty problems in a 50-minute timespan. Ten of the questions are in English.	Five major question problems during 120 min, and one of these appears in English.
Methods	Students are required to interpret geographical knowledge and concepts from photographs and charts that are projected onto a screen for a duration of 1 min.	Students are required to answer both the Japanese and English questions in written essay form.
	Students are required to match the given information against four possible answers for each question.	Students are required to use a variety of materials, such as maps and statistics, for a broad perspective of their abilities in thinking, making judgments, and expression, based on such knowledge and concepts as exploring the background context and causal factors, identifying social issues, and proposing solutions.

Also, experts and exchange students from the host country provide information on local geographic conditions, giving Japanese students deeper insight into the characteristics they will encounter.

Since 2011, the JST has taken on the role to administer the call for contestants in the Geography Olympiad, along with the other Science Olympiads in order to attract more exam takers for the Japanese championship. To ensure that administrative work proceeds efficiently, the Executive Committee was formed in 2012 to strengthen its business office system. Information about the competition is disseminated to all high schools in the country and teachers responsible for geography in their schools are notified, with attention paid to the Super Science High Schools (SSH) designated by MEXT. These strategies have resulted in a rapid increase in the number of exam takers, a contributing factor to the revitalization of geography education.

The exams are written at a high level of difficulty, and English ability is necessary for participation in the international competition. This restricts contestants in advanced schools (e.g., Super Science High Schools), as a way to prevent a concentration of competitors with elite geography opportunities. To broaden the pool of potential contestants and to foster equal opportunity for students in all schools settings, the Executive Committee has changed the questions so that 80 % are in Japanese. Additional mechanisms are needed to attract more contestants. Promotion of the Geography Olympiad is one way to revitalize geography education in Japan, and is an opportunity for the exam takers to significantly upgrade their geography skills.

13.3.2 *Exhibition of Children's Environmental Maps*

The Exhibition of Children's Environmental Maps ("Maps of Our Surrounding Environment") is an event put on by the SEME, and is based in Hokkaido. Many geography-related research organizations (e.g., AJG), research institutes (e.g., the Geospatial Information Authority of Japan (GSI) and National Institute for Environmental Studies (NIES), local governments and education committees, and map-based enterprises (e.g., publishers and mass media) support or encourage this event. The children's maps are submitted from all over the country and even from overseas, reaching about 1,500 submissions annually. Similar exhibitions are also held in other areas of the country (Masuda 2008). However, in those cases, map entries are limited to the local area. There is no other exhibition that receives submissions from both all of Japan and abroad equally. The Children's Environmental Map Contest is the largest map exhibition of any kind in the country (Himiyama 2012).

Since it was first held in 1991, the Children's Environmental Map Contest has been held annually in late October, and will have been held for the 23rd time in 2013. The objective of the exhibition is to foster children's interest and concern for community problems through map making. At the same time, children have the chance to develop talents that can contribute to improving environmental harmony in the local region.

Submissions come primarily from individual schools, most of which include the map creation process as part of study units on the local area or study units on maps. All maps from both Japan and other countries are judged rigorously by the Executive Committee. The maps are to depict children's observations of their environment. The theme of 2013 was "Hazard prevention" but children could also choose to map a theme of their choice. Generally, many maps depict skillful creations that display geographical patterns and contemporary issues.

The competition rules stipulate to "Make a map of the environment close to you that shows the things you surveyed, observed, and thought about". Children's creativity is valued and left up to the map makers. Maps that do not fit a traditional map type but rather show a high degree of originality are usually highly evaluated.

The process of map making includes the following steps: (1) Decide on the map theme and the limits of the area to be surveyed; (2) conduct diligent surveys and observations in the field; (3) organize and map the survey data; (4) analyze and interpret patterns in the finished maps; and (5) communicate the conclusions. The goal of the Children's Environmental Map Exhibition is to nurture geography skills in area surveys.

The new National Curriculum Standards firmly implements the fundamentals of exploratory and language activities and emphasizes these as pillars of learning. This supports geography education's position to provide students with opportunities to learn mapping and area survey skills, which are practices related to problem-solving abilities. In the same way as the Geography Olympiad, the Exhibition Children's Environmental Map Contest reinforces scholastic abilities called for in the new National Curriculum Standards.

13.4 Examples of Recent Outstanding Social Action Programs: In-service Training of Teachers

13.4.1 *Efforts by AJG-CGE*

Although not regularly scheduled, the AJG-CGE has been holding training sessions for elementary and junior high school teachers since 2007. These in-service training sessions are organized along themes that use geography skills such as "Methods for Using Atlases with Literary Works" and "Constructing Thematic Maps and Distribution Maps with Mapping Software." This service is necessary because many teachers licensed to teach geography lack these disciplinary skills (Akimoto et al. 2010). For example, teachers in elementary and junior high schools are not required to major in geography at university. Particularly in the case of elementary school teachers who instruct all subjects, there are many for whom Social Studies is outside their area of expertise, let alone Geography.

Students in third and fourth year elementary school are expected, in their Social Studies class, to about their region at the level of cities, towns, villages, and prefectures, and geography skills of map reading and map making. Junior high school students take a geography class that includes mandatory activities such as surveying nearby surroundings, an exercise that applies skills unique to geography (e.g., field observation, reading topographic maps, making thematic maps, and writing survey reports). However, as mentioned, most elementary and junior high school teachers have not majored in geography and have not necessarily systematically learned geography skills. For this reason, there are many cases in which instructors are unable to lead effective geographic lessons and field research of the local area.

In Japan, there is a tendency for curricula to emphasize knowledge of facts over knowledge of methods. As evidence, course content in geography skills is not systematically specified and, although the cognitive development of students is discussed at academic conferences and research meetings, it is rarely examined in the schools.

If during teachers' training they learn about geography curricula that systematically outline how to instill geography skills, this could substitute for their gaps in geography knowledge. However, this is insufficient given the realities of students need to succeed in society. Teacher candidates are so poorly prepared in geography that it further complicates the situation. For example, elementary pre-service teachers need only take two required credits in "instruction methods for subjects that are related to Social Studies" to qualify as a teacher. For junior high school teachers specializing in Social Studies, credits in general geography and regional geography, which are considered "courses related to the subject," are required. But the content of these is broad, covering everything from history and civics to geography education. Since Geography has become an elective subject at the senior high school level, most of the teachers did not themselves take Geography when they were in senior high school, thus many of them are standing at the teacher's podium with only the knowledge of a junior high school level of knowledge (Iwamoto 2006).

In that light, teachers' acquisition of geography skills is important and the in-service training sessions implemented by AJG-CGE are thus extremely significant. Participating teachers are very appreciative and provide feedback such as "I can use this in tomorrow's class." Plans for the future to tie together these training sessions with the teacher's license renewal courses offered at universities and prefectural education committees would help revitalize geography instruction in the classroom, and create opportunities for many teachers to participate in the training.

13.4.2 In-service Training of Teachers Implemented by Other Associations

The HGSJ established a Geography Education Research Sub-group in 2005, and somewhat belatedly has become involved organizationally in the revitalization of geography education. Their Geography Education Summer Training Sessions have become one of the major cornerstones of the sub-group's activities. These have been held every year since 2005 during the schools' summer vacation and are designed for non-members of the association, such as elementary, junior high and senior high school teachers and university students who are planning to become teachers. The first training session was held in August 2005 in Kyoto's Tanabe City Community Hall and the surrounding area, with an overall theme of "Turning the Example of the Local Area into Teaching Material." During the morning, field trips were held, and in the afternoons lectures and reports on examples of practice in geography education were given (Kobashi 2007). Until 2011, the sessions were held at different venues across the prefectures but primarily in the Kinki District. The sessions each attracted about 50 participants annually (75 at maximum).

13.5 Conclusion and Remaining Issues

In this chapter, on the authors have described a range of formal and informal geography education outreach activities. These include the education activities of the AJG, along with an overview of their social action programs since the 1980s, described synopsis of the Geography Olympiad Japan Championship, the Children's Environmental Map Contest, and in-service training sessions for teachers. These are outstanding recent examples of social action programs related to geography education, and their implementation process, results, and issues that have been discussed. A brief overview of geography education efforts from other academic associations such as the HGSJ is also presented for comparison purposes.

Some have considered geography education at the senior high school level to be in a precarious state. The reduction in geography enrollments, geography as an elective rather than compulsory course in senior high school, and the fact that it is no longer one of the university entrance examination subjects, have all contributed to the decline of geography education. This situation has also been affected by political debates in the education committees of Kanagawa Prefecture, Yokohama City, and Tokyo Metropolitan district, who have pushed for a policy making Japanese History a compulsory subject in high school, to foster Japanese who can live as major actors within international society. As a result, high school students in these cities and prefectures have been robbed of the opportunity to take geography as an elective, owing to a lack of class hours. This results in student populations who will not have a balanced education, one that focuses more on history than geography.

Globalization makes geography an indispensable subject, especially the skills and perspectives to evaluate the past, focus on the present, and to envision the future. In order to instill these abilities, students need to be aware of the world in both its temporal and spatial dimensions (Terao 2008), the raison d'être of senior high school geography and history subjects. We believe that Geography is in a perilous position because it is equated with simply learning place names, and the field does not do more to educate the public of geography's connection to society. As long as geography education's social contributions are visible, the future of geography education seems slim.

From the viewpoint of forming an image of the world, responding to various contemporary issues, and contributing to active participation in society, the role of geography education in the new National Curriculum Standards clearly defines these goals. That is to say, as the various contemporary issues (e.g., of the environment and development, disasters and disaster prevention, ethnicity and religion, resources and energy, population and food supply, and others) continue to intensify across wide areas of the world from local to global scales, the educational world collectively demands policies of response and social contribution from geography education.

In what ways will students apply their geography skills and abilities to solve local, national, and global issues? How should they create a better society for the future, given the increasing complex world? What is now being asked of geography education is that it nurtures citizens who will have the competencies to participate in society as major actors.

References

Akimoto H, Takizawa Y, Ishizuka K, Hirasawa K, Agemura Y, Komiya M (2010) Shogakkou kyouin yousei ni okeru chiri kyouiku no genjou to kadai: shinki saiyou kyouin he no ankeeto chousa ni yoru bunseki (The present conditions and problems of the geography education in the teacher-training of elementary schools: an analysis by the questionnaire survey for newly-appointed teachers). Shin Chiri (New Geogr) 58(1):33–42

Himiyama Y (2012) Kankyou chizu sakuhinten no hatasu yakuwari to igi (The role played by the environmental maps exhibition and its significance"). Chiri (Geogr) 57(7):12–17

Iwamoto, H. (2006). Kyouin yousei daigaku no chiri kyouiku (Geography education of teacher training universities). In: The Geographic Education Society of Japan (ed) Chiri kyouiku yougo ginou jiten (The dictionary of geographical terms and skills). Teikoku shoin, Tokyo, p 65

Izumi T (2013) Kokusai chiri olympic de tamesareru 'chiriteki nouryoku towa: iGeo Kyoto taikai he mukete (What is geographical ability which is necessary in international geography Olympiad?). Chiri (Geogr) 58(7):12–19

Kobashi T (2007) Jinbun chiri gakkai no chiri kyouiku kaki kenshuukai (The human geographical society of Japan's summer training session). Chiri (Geogr) 52(7):22–26

Kohno K (1996) Koutou gakkou ni okeru 'chiri' no rishuugen: kyoukasho saitaku busuu no suii yori (Decline in enrollment in senior high school geography). Chirigaku Hyoron (Geogr Rev Jpn) 69A(12):966–967

Masuda S (2008) 11 kaime wo mukaeta jidou seito chizu yuushuu sakuhinten (Looking towards the eleventh exhibition of prize-winning maps by children and school pupils). Chizu Chushin (Map Center) 426:8–9

Shimura T (2007) Nihon chiri gakkai no taikaiji ni kaisai sareru 'chiri kyouiku koukai kouza' (The geography education open public courses held at the time of the Association of Japanese Geographers' conference). Chiri (Geogr) 52(7):20–21

Terao T (2008) Nihonshi jugyou no naka no chizu riyou– nihonshi to chiri no setten wo kangaeru– (The use of map in japanese history teaching: a point of contact between Japanese history and geography). Chiri Kyouiku Kenkyu (J Geogr Edu) 1:74–78

Chapter 14
GIS and Geography Education in Japan

Tomoaki Ito

Abstract Geographic Information System (GIS) has been included for the first time in Japan's revised National Curriculum Standards. What is required to implement GIS in Japanese geography education? How do teachers use GIS to meet the aims set out in the National Curriculum Standards? The National Curriculum Standards place great importance on GIS in senior high school geography classes, requiring that students use this technology throughout their geography education to develop relevant skills in geography. GIS is actively included in the new National Curriculum Standards for senior high schools, more so than in other subjects (such as Information classes in senior high school, Social Studies in junior high school, and others). However, the teaching of GIS is not fully implemented because many problems exist. For example, teachers need to recognize the value of GIS in imparting geographic skills and they need to be part of a community that shares best teaching methods and lesson plans. An environment that enriches geography teaching through GIS use is of great importance.

Keywords GIS • New National Curriculum Standards • Senior high school • Sharing • Teaching materials

14.1 Introduction

The discussion to include geographic information systems (GIS) in geography education has been ongoing since the late 1990s. The new national curriculum of 2009, implemented beginning in 2013, stipulates that GIS is to be taught in senior high school geography. This represents a turning point in Japan. GIS has transitioned from being in an exploratory phase to a point where best implementation practices are being discussed. However, teacher interest in adapting GIS to the classroom is low, and it is difficult to determine whether GIS has been sufficiently integrated into

T. Ito (✉)
Shizuoka Prefectural Susono Senior High School,
900-1 Sano, Susono, Shizuoka 410-1118, Japan
e-mail: geo-ito@bea.hi-ho.ne.jp

Y. Ida et al. (eds.), *Geography Education in Japan*, International
Perspectives in Geography 3, DOI 10.1007/978-4-431-54953-6_14

geography education. In addition to GIS knowledge, it is important to discuss the impact of GIS on geography education and why it is to be promoted as a tool.

Within this context, the author discusses the position of GIS in Japanese geography education, focusing in particular on senior high school. First, an overview of the ongoing discussion concerning implementation of GIS is presented. Second, the role of GIS as laid out in the new national curriculum is examined. This is followed by a comparison of GIS education across senior high school Geography A, Geography B, and other subjects in which GIS is taught. Finally, the barriers to implementing GIS in the senior high school education curriculum are clarified, and by suggesting resolutions to these problems, this chapter delineates the ongoing role of GIS in geography education.

14.2 Diffusion Process of GIS in Japanese Geography Education

Deliberation on the integration of GIS in geography education has been ongoing since the 1990s. University researchers were the first to suggest introducing GIS into school geography classes. This discussion began by examining GIS education in the United States as a model. The history of GIS development and its spread across the U.S. has resulted in its role in geography education (e.g., Usui 1993). Discussion in Japan led Akimoto (1996) to suggest the idea of including GIS in Japanese senior high school education. The importance of having a curriculum and standards for this educational tool was also established (e.g., Itoh et al. 1997). Other scholars noted the advantages of GIS for the selection and comparison of data, leading to changes to pedagogy in the seventh to twelfth grades that moved away from a lecture-style teaching method to a process focused on individual students making personal choices using technology (Ida 1999; Sugimori 1999; Akimoto 2003a).

Since the 2000s, university researchers and school teachers have developed GIS teaching materials that are fit for problem solving in a computer lab setting. GIS is currently viewed simply as class work where instructors teach students how to use GIS software, and they use the software to solve specific problems (Ito 2005). It is uncommon that GIS education is seen as a supplementary tool in the regular classroom environment.

Teachers face a number of obstacles to integrate GIS in the classroom. For one, the cost of computer equipment and software poses affordability issues (Minamino 2003; Koseki 2008). The financial burden delayed the mention of GIS in the national curriculum until the most recent version, to take effect beginning in 2013 (Neda et al. 2008). A study on junior and senior high schools in Hyogo Prefecture found a paucity of available GIS-related teaching materials and methods published in academic literature or by teacher associations (Kobashi 2007), making it more challenging for teachers to find GIS resources.

Inclusion of GIS in the new National Curriculum Standards has been a turning point for both GIS research and geography education. For this curriculum reform to sustain GIS in the education system, issues regarding hardware, data availability, and teaching resources as well as best practices and educational research are needed.

14.3 The Position of GIS in the New National Curriculum Standards

This section discusses the position of GIS in geography education as mandated in the new national curriculum, as well as its impacts. Here the focuses on how GIS can be implemented in the senior high school subjects Geography A and Geography B. By contrasting this with GIS in junior high school Social Studies and related subjects in senior high school, the content of geography taught using GIS will be clarified.

14.3.1 The Position of GIS in Senior High School Geography Education

According to the new National Curriculum Standards for senior high school geography, the use of GIS is to be considered for all applicable content of the subject. However, there are different expectations between Social Studies in junior high school and related senior high school subjects. For Geography A and Geography B in senior high school, the curriculum maintains that in order for students to obtain a geographic way of thinking, to look at phenomena, develop the ability to read and construct maps, as well as to interpret satellite maps and aerial photos, students should be instructed in a planned and systematic manner. In that regard, as well as being able to use teaching materials (maps) effectively, it is desirable that "students are able to use information and telecommunication networks and geographic information systems (GIS) in order to effectively collect and analyze geographic data like maps and statistics" (MEXT 2009, p. 27).

MEXT (2010) notes that acquiring geographic skills is not just the ability to read and draw maps. It includes the ability to read geographic data from maps as well as to take geographic data and express it in map form. This goal can be achieved by using digitized geographic data and analysis with GIS. In addition to deepening geographic knowledge and skills, GIS can be effective in supporting student information use acquisition.

According to the latest National Curriculum Standards, "geographic skills" equates to acquiring geographic data from maps, and using geographic data to construct maps. GIS is well positioned to support effective acquisition of those skills. The development of geographic skills is the fundamental purpose of geography education as taught in Social Studies and Geography at the junior high to senior high levels, respectively. For example, the new National Curriculum Standards mandate that junior high level Social Studies must encourage the application of geographic information to develop basic geographic skills. "Geographic skills" are defined as "skills necessary to apply geographic information" and "skills in using maps," and these are the basis of GIS use (Tatsuoka 2002). GIS education is built on the principle that GIS and geographic information sciences can effectively teach research-based activities, especially in Social Studies education (Sawamura and Nagase 2005). Given these arguments, GIS should not be limited to geography in senior high school, where it is currently stipulated, but it should also be systematically and progressively included within related subjects to further develop students' geographic skills.

According to the Geography Annex of the National Curriculum Guidelines, it is imperative that GIS lessons are designed in conjunction with geography and related subjects, particularly Information courses, for the most effective use of GIS equipment and hardware (MEXT 2010). Thus, appropriate facilities and equipment as well as teacher training are important for implementing GIS-based education.

14.3.2 GIS-Related Subjects in High School Other Than Geography

In addition to Geography A and B, GIS is taught in other closely related subjects, such as the compulsory high school subject Information. As previously mentioned, the new national curriculum envisions GIS as a means to foster strong links between Geography A and B with Information. Synergy between these two subjects can be easily seen within the software and hardware elements of GIS, in addition to its analytical capabilities (Akimoto 2003b).

In most senior high schools, Information is a compulsory subject in the first year whereas Geography A and B are offered as elective courses from the second year onward. It is possible that students who take information studies classes and learn GIS will go on to choose geography the following year. Thus, including GIS within Information is a possible way to increase the number of students who take geography as an elective subject.

GIS is mandated in various courses at vocational schools, such as the technology school subject of "Surveying", the agricultural school subject "Management of agricultural information", and in marketing courses in commercial schools. In these cases, students learn to apply GIS to their profession. In contrast to the academic stream of senior high school, students learn GIS theory and technological functions. GIS is not recommended to be taught throughout Geography A and B courses.

Although GIS instruction should be actively encouraged in these courses, only one teaching unit is credited to include the principles and methods of these core subject, unlike Geography A and B courses that introduce GIS education.

14.4 Strategies for Dissemination of GIS in Geography Education

The new national curriculum dictates active implementation of GIS, especially in senior high school geography education.

The value of GIS lies in its contribution to society. GIS has transitioned from a geography information "system" to a "science," and it is predicted that this will move toward a focus on "service" (Yano 2001). Geographic information has become a part of our everyday lives and through a variety of media, its use has changed the manner in which business is conducted. The inclusion of GIS in the National Curriculum Standards, with a focus on senior high school geography, reflects recent changes in society.

However, the diffusion of GIS in education is confronted by barriers. One obstacle is that geography teachers do not have equal access to computers. Instead, subjects like Information have first priority for computer use. The second issue is that teachers inexperienced in GIS have difficulty developing resources and materials for lectures and interactive exercises. Teachers need standardized knowledge of information and communication technologies (ICT) as well as concepts unique to GIS (e.g., how to save data and the file formats).

The following section provides possible solutions and policies that may help to promote GIS in senior high school geography education.

14.4.1 Lesson Plans Using Textbook-Based 50-Minute Lessons

Ito (2010) introduced the concept of virtually free, textbook-based 50-minute lesson plans to be used as GIS teaching materials. Developed with a modest budget, these resources encourage use of free GIS software and statistical data within a 50-minute period. These lessons move students from constructing maps to the activities of comparison, analysis, and observation, to acquire the knowledge and skills required in geography studies.

Data are cleaned and ready for immediate use, so that students do not need to gather nor spend many hours processing data, which are time-intensive tasks. With endless application opportunities that is tied back to the teaching unit, students become interested in manipulating data with GIS software and look forward to their next opportunity for such practical work. Through active visual recreation or data manipulation, such as classifying data based on a set of specific conditions (e.g., the top ten countries or countries below an average value in a category), students begin to create individual maps and interpret the mapped data. Through these exercises, they become proficient at summarize and communicate their results. Each successive class builds on knowledge and skills learned from previous activities in a single unit of a 50-minute period lesson. In contrast to Information courses that involve multiple teachers or teaching assistants, practical work for GIS generally requires only one teacher who can lead up to 40 students in a class. Generally, the teacher demonstrates steps and ideas for the activity, which students observe before doing the hands-on lesson themselves. The author suggests teaching GIS in 50-minute blocks completed by student reading of textbook lessons as a way to focus GIS learning.

14.4.2 Development and Provision of Frequently Used Teaching Materials

In practice, even for teachers who are able to effectively use ICT, it takes time to become well acquainted with GIS software. In 2006, the author received a research grant from the government technical skills promotion research association to study and develop versatile teaching materials using free or low cost GIS software. The research website (www.itochiri.jp) and blogs have been accessed more than 200,000 times within a 1-year period. Teaching resources and data files are free to download. These are compatible with "MANDARA," the most widely used free software in Japan (.mdr format), or with Google Earth (.kml and .kmz formats). If the materials were organized and disseminated via teaching association or textbook company websites, more rapid integration of GIS into geography education would be likely.

To buttress GIS instruction and to make it a commonly taught element of geography, teachers have an important role to play. Examples of teacher activities that can help promote GIS include sharing resources with fellow teachers on a local or broader scale, and organizing working groups to develop novel instructional materials, with advice from GIS experts.

14.4.3 Necessary Equipment and Effective Use of Teaching Aids

Like other subjects, GIS instruction is affected by limited resources, which has a direct impact on facilities and equipment that are necessary to teach GIS within geography.

School teachers usually have little experience setting up technical equipment, in addition to time limitations. It is therefore necessary to approach experienced university researchers and administrators for support. Cooperation between such interested stakeholders would be valuable for promoting GIS in senior high school geography education.

14.5 Conclusion

This chapter began by discussing the current use of GIS in geography education, with a particular focus on its place in senior high schools. The new National Curriculum Standards mandates active GIS implementation in senior high school geography classes. Through synergistic activities with Information and other geography-related subjects, it is possible to create appropriate GIS teaching materials and methods.

According to the National Curriculum Standards, the ability to effectively use geographic information and the maps are to be a focus of geography at the junior high through senior high school levels (Shimura 2010). However, it should be emphasized that the ability to effectively use geography information is not limited to GIS. For students to acquire these skills, maps and geography textbooks are useful teaching materials as well.

GIS effectively demonstrates data manipulation and management of large datasets to foster an understanding of geographic information that would not have been possible without technology. In addition, students' independent use of computers to learn concepts and to solve problems is an important educational experience. GIS knowledge and skills are applicable across subjects. While developing synergy with related subjects, it is helpful to illustrate how different skills (e.g., using geographic data, using the inquiry process) are transferable to these subjects.

Geography learning can be enriched through developing GIS resources, supporting leaders in GIS education, and building capacity within the geography community. Equally important is the need to support and work closely with the community of educators and researchers who are actively involved in strengthening GIS instruction in classrooms.

References

Akimoto H (1996) GIS to koko chirikyoiku (GIS and high school geography education). Shin Chiri (New Geogr) 44(3):24–32

Akimoto H (2003a) Chuto chirikyoiku ni okeru GIS no igi (The role of GIS in secondary geographical education). GIS-Riron to Ouyou (Theor Appl GIS) 11(1):109–115

Akimoto H (2003b) Koto gakko futsu kyoka "Joho" niokeru GIS no igi to kanosei (The role of GIS in high school information education). Joho-kagaku Kenkyu (Inform Sci Res Dokkyo Univ) 21:31–40

Ida Y (1999) Chiri kyouiku ni okeru ishi kettei tool tositeno GIS (Using GIS as the tool for decision making in school geography). Chiri Joho system gakkai kouen ronbun-shu (Conf Abstract GIS Assoc Jpn) 8:7–8

Ito T (2005) Seito ni teiji suru kyozai niyoru chikeizu no dokuzu shien (The practical report of topographical map reading with "show at once GIS"). Shin Chiri (New Geogr) 52(4):32–43

Ito T (2010) Itochiri shiki: chiri no jugyo ni GIS (The Itochiri methods: GIS for geography classes). Kokon shoin, Tokyo

Itoh S, Ida Y, Nakamura Y (1997) Gakko kyoiku ni okeru GIS riyo: America gashukoku no doko to wagakunino kanousei (The use of GIS in school education: its trends in the United States and possibilities in Japan). GIS-Riron to Ouyou (Theor Appl GIS) 6(2):65–70

Kobashi T (2007) Hyogo ken ni okeru kyoiku GIS jissen (GIS lessons in Hyogo prefecture). Hyogo Chiri (Hyogo Geogr J) 52:11–18

Koseki Y (2008) Kyoiku GIS no jissen to kongo no tenbo (The practice and prospects of GIS in education). Chiri Shisou (Ann Geogr) 50(1):138–147

MEXT (2009) Kotogakko gakushu shido yoryo (National curriculum for senior high school). Higashiyama Shobo, Kyoto
MEXT (2010) Kotogakko gakushu shido yoryo kaisetsu: Chiri-rekishi hen (National curriculum for senior high school-explanation-: geography and history). Kyoiku Shuppan, Tokyo
Minamino T (2003) Wagakuni no gakko kyoiku niokeru GIS katsuyo no genjo to kadai (GIS in school education, Japan; a review and prospective). Chiri-Kagaku (Geogr Sci) 58:268–281
Neda K, Kokudai Y, Murakami F (2008) Kotogakko chiri A no kyokasho ni sokusita GIS jugyo jissenn (Case study of practicing GIS education according to a textbook at high school). Kyoiku Jissen Sougo Center Kenkyu Kiyou (Bull Center Edu Res Dev Nara Edu Univer) 17:201–206
Sawamura K, Nagase H (2005) Joho katsuyo noryoku wo ikusei suru chugakko shakaika no kyozaikaihatsu (Development of teaching materials for junior high school social studies to support the use of information on capacity-building). Nihon kyouikukougakkai ronbun-shi (Proc Jpn Soc Edu Technol Res) 5(6):31–36
Shimura T (2010) Shin gakusyuu shidou youryou no syushi wo fumaeta Chizu gakusyuu no arikata (Prostects for map teaching and leraning practices under new courses of study in Japanese national curriculum 2008/2009). Chizu (Map J Jpn Cartogr Assoc) 48(2):19–28
Sugimori K (1999) Chiriteki na shikoryoku wo takameru tameno kyoiku GIS nitsuiteno ichi kosatsu (A study on the educational GIS for developing a geographical thinking). Chiri Joho system gakkai kouen ronbun-shu (Confer Abstract GIS Assoc Jpn) 8:27–30
Tatsuoka Y (2002) Shakaika kyoiku no tameno GIS enyo no tameni? teikakaku application no kanosei(For applying GIS to social studies : a case for low-cost applications). Kyoiku Jissenngaku Ronsyu (J Sci School) 3:79–88
Usui T (1993) Chiri joho system kenkyu to GIS kyoiku no hitsuyosei (The necessity of GIS research and GIS education in geography). Nara Daigaku kiyou (Bull Nara Univ) 21(1):57–65
Yano K (2001) GIS, cyber space no mizusaki annainin (GIS, a navigator of cyber space). Jinbun Chiri (Jpn J Hum Geogr) 53(5):50–55

Chapter 15
GIS Specialists' Support for Geography Education

Hideki Oshima

Abstract Senior high school teachers in Japan are obliged to develop practical applications of geographic information systems (GIS) to meet the new National Curriculum Standards, although they have little formal education in GIS during their undergraduate and in-service training. Therefore, GIS specialists have supported the schools in choosing suitable GIS for their facilities and budgets, in training teachers to manage GIS, and in the development of GIS teaching materials. These experiences revealed that: (1) Reading GIS maps is a higher priority than editing them; (2) delivering datasets via web GIS is the easiest method, as long as the school internet can support the bandwidth; and (3) GIS training videos are required to demonstrate how to create GIS maps in a classroom without the internet.

Keywords Geography education • GIS • GIS movies • Senior high school • Web GIS

15.1 Introduction

Senior high school geography teachers in Japan must develop practical geographic information system (GIS) use to meet the new National Curriculum Standards. However, they have little formal education in GIS during their undergraduate and in-service training, because many of their professors also had little GIS knowledge. Although some teachers have learned GIS through other means, they continue to face difficulties such as slow internet speed at their schools, which affects web GIS use. Another issue is the compatibility of saved maps when shared across computers. Given these limitations, GIS specialists have been hired to support schools to resolve these problems. This chapter reviews the roles of these specialists and how this effort is supporting the spread of GIS use in schools.

H. Oshima (✉)
Keio University, 301 Annex Hiyoshi, 3-18-20 Hiyoshi, Kohoku-ku, Yokohama 223-0061, Japan
e-mail: oshimapanda@gmail.com

Y. Ida et al. (eds.), *Geography Education in Japan*, International Perspectives in Geography 3, DOI 10.1007/978-4-431-54953-6_15

15.2 Support Menus by GIS Specialists

A survey was conducted with senior high school teachers regarding barriers to implementing GIS in class lessons (Fukuda and Tani 2003). Teachers noted a lack of computer facilities, budgets to buy GIS software, in-service teacher training, and textbooks. Ito (2012) suggested that GIS could be taught across multiple subjects. If teaching resources and lessons were shared between teachers, support would be more accessible. For optimal GIS instruction, Itoh (2004) described a list of must-haves: GIS software and websites, and in-service training for GIS learning.

15.3 Choosing and Developing Suitable GIS

Teaching styles, available facilities, and budgets are requirements for determining a suitable GIS at each school. Table 15.1 summarizes GIS software and functions already in use in Japanese schools.

15.3.1 Requirements for Suitable GIS

In a linguistics context, the ability to read sentences is a requirement to write them. The parallel in geography education is that the ability to read maps is the foundation to edit them. Various types of GIS can be used with map data for visualization and

Table 15.1 Functions, facilities and budgets for each GIS

Software	Web GIS	Google Maps/ Earth	Picasa	Chizutaro	MANDARA	Quantum GIS	Arc GIS	Green Map/ Hi map mister
Functions								
Reading GIS map								
Topographical map	•			•	•			
Town map	•	•	•	•	•	•	•	
Aerial photo	•	•	•		•	•	•	
Editing GIS map								
Editing GIS data	•	•		•	•	•	•	
Editing photo	•		•	•				
Statistic map	•			•	•	•	•	•
Elevation map	•			•	•	•	•	
Facilities								
Offline use				•	•	•	•	•
Budgets								
Free	•	•	•		•	•	•	

editing. Some GIS maps are easily read but cannot be edited, whereas others can be easy to edit but difficult to interpret without advanced skills. Determinants of GIS selection depend on teaching facilities and budget.

The author currently operates the website "Easy Digital Maps[1]" to introduce "easy" GIS for beginners. The materials are categorized in the following groups: (1) "Looking at GIS Maps", (2) "Looking at Old GIS Maps", (3) "Looking at Satellite Data", (4) "Looking at Maps in an Atlas in GIS", (5) "Looking at Statistics for Each Street Number of Districts in a Town", (6) "Looking at Local Disaster Risks", (7) "Looking at Distribution of Stores", and (8) "Editing GIS Maps". A record of webpage visitor logs was analyzed for one full year in 2010. The findings indicate that only 20 % of all visitors viewed "Editing GIS Maps". Thus, most GIS beginners simply want to look at GIS maps rather than edit them. This trend applies to both teachers and student users.

The author found that in addition to perusing and editing maps, teachers mark on printed GIS maps. They select visible items (e.g., roads or buildings), adjust the scale and zoom the map with GIS, and then print the outline map. Students also mark symbols or other features on the printed map. Used in this way, GIS is a platform for looking at maps, loosely coined "analog GIS".

GIS functions can be used for all geographic instruction. For example, reading GIS maps can be satisfied by displaying topographic, large-scale, and satellite maps, as well as aerial photographs. In case students are expected to make GIS maps, they can manipulate data to create thematic maps.

The other requirements for implementing GIS in schools are facilities and budget. Money is needed to set up the facility, which includes but is not limited to personal computers, software, and internet access in the classrooms. Fujisawa City (2007) reported that a GIS specialist advised teachers in the city to use the desktop GIS software Chizutaro. As a result, it was installed on all personal computers for teachers in every primary and junior high school in the city. Other sources of suggestions include the Ministry of Land, Infrastructure, Transport and Tourism (MLIT),[2] which introduces GIS and selection requirements in their teacher manuals.

15.3.2 *Selected and Developed GIS*

15.3.2.1 Web GIS

Web GIS uses a web platform by storing data in the cloud to support GIS activities and only requires access to the Internet and a web browser. Web GIS has the capability to debug issues quickly and revise online programs, because the application is on the provider's server(s). Datasets are also stored on that server, allowing the user immediate access and use. Ugawa et al. (2003) saw web GIS as a solution to current budget cuts, simplifying operation menus, and free sharing of maps without a need

[1] https://sites.google.com/site/dejichizu/. Accessed 31 July 2012. (J)

[2] http://www.mlit.go.jp/kokudoseisaku/gis/gis/kyoiku/index.html Accessed 31 July 2012. (J)

to download or purchase software. Datasets served on web GIS are the easiest means of school access, given that internet speeds exceed 30 mbps.

Many schools in Japan are facing budget deficits and do not have funds for internet connection. A classroom internet (LAN) for all schools in the same city is connected to the city education center. At least one internet circuit is connected to that center. To prevent students from browsing non-educational sites, software is used to block search engines such as Google and Yahoo. This filter makes it difficult to access web GIS sites, because Google Maps[3] is one mapping platform used in classrooms. Its use during instruction is for the display of elements on town maps, as well as satellite and aerial photos. With a login account, students can have editing capability with the "My Places" menu.

Web GIS sites for geography education are developed by GIS specialists. For example, Murayama (2002) developed "Web GIS for Geography Education,[4]" which stores census statistics on population, industry, and cultural information for cities, prefectures, and nations. Users can create and edit maps, produce scatter plots, and do multivariate analysis. Another platform was designed by Ugawa et al. (2003) called "Let's inspect it—The frog map".[5] This website stores base maps and allows users to browse and edit. There is an element of citizen science, in which students can input frog habitats on a public web map. The last example is led by Oshima et al. (2011), who operates "Teaching materials for 'geographic and earth scientific experiments' in junior and senior high schools with satellite data".[6] This site uses a web GIS management software (ArcGIS Server[7]) and stores satellite images (visible and infrared), population density maps, green coverage maps, and heat island maps for browsing.

15.3.2.2 Desktop GIS Software

Desktop GIS software allows users to display and edit spatial data stored in personal computers or on servers. Data can be accessed and manipulated online or offline. However, there is a financial cost to purchase the software and robust computer hardware is needed. Fortunately, there are some free desktop GIS software that have a low-demand computer specification. Generally, proprietary desktop GIS software has advanced and additional functions relative to web GIS. Some desktop GIS software has been revised for geography education. A summary of these and associated software are provided below.

---

[3] https://maps.google.com/maps. Accessed 31 July 2012.

[4] http://giswin.geo.tsukuba.ac.jp/teacher/murayama/edugis/index.html. Accessed 31 July 2012. (J)

[5] http://map.edb.miyakyo-u.ac.jp/kaeru/. Accessed 31 July 2012. (J)

[6] http://earthgis1.isc.chubu.ac.jp/Jikken/. Accessed 31 July 2012. (J)

[7] http://www.esri.com/software/arcgis/arcgisserver. Accessed 31 July 2012.

MANDARA[8] Tani et al. (2002) revised this free software for junior high school student use, from the original that was aimed at university students. MANDARA can display topographic maps, town maps, and aerial photos, and support color classification of statistical maps with Excel data. Tani et al. (2002) added a "Start Menu" and a form for students to input property data.

Chizutaro[9] This software has been revised to accommodate minor requests from primary and junior high school teachers in Fujisawa City (2007). The original Chizutaro was a desktop GIS software for local government use. It displays topographic maps and town maps, has an edit function for data and photos, plus a color classification option for statistical maps. Its cost is attractive for school implementation, only 3,500 yen (approximately 35.00 USD).

Google Earth[10] This is a free digital globe software that can display large-scale maps, satellite and aerial photos, and has an edit feature.

Picasa[11] This is a free photo software that displays photos in Google Maps.

Kashmir 3D[12] This is a free GIS software that displays topographic maps.

Quantum GIS (QGIS)[13] This is an open-source GIS that can display topographic maps, town maps, satellite and aerial photos, and allows editing of GIS data and color classification of statistical maps.

ArcGIS Desktop[14] This is an international commercial GIS software. Schools in Japan have a 2-year license with free support for primary, junior, and senior high school education. Since ArcGIS is a specialized and complex system, trained personnel are needed to manage and install it on computers. It has the same functions as QGIS.

Textbook Publishers School textbook publishers provide software that displays maps from atlases and supports color classification of statistical maps, such as "Hi Map Mister[15]" and "Green Map".[16] Diercke (2010) is a world atlas written in English, published in Germany. A license key is provided in the textbook to access the digital globe software "Diercke Globe".[17]

---

[8] http://ktgis.net/mandara/index.php. Accessed 31 July 2012. (J)

[9] http://www.tcgmap.jp/product/chizutaro/. Accessed 31 July 2012. (J)

[10] http://www.google.com/intl/en/earth/index.html. Accessed 31 July 2012.

[11] http://picasa.google.com/intl/en/. Accessed 31 July 2012.

[12] http://www.kashmir3d.com/. Accessed 31 July 2012. (J)

[13] http://qgis.org/. Accessed 31 July 2012.

[14] http://www.esri.com/software/arcgis/arcgis-for-desktop. Accessed 31 July 2012.

[15] http://www.teikokushoin.co.jp/products/pc_soft/index01.html. Accessed 31 July 2012. (J)

[16] http://www.tcgmap.jp/product/greenmap_world/. Accessed 31 July 2012. (J)

[17] http://www.diercke.com/dierckeglobe_download.xtp. Accessed 31 July 2012.

15.3.2.3 Datasets for Desktop GIS Software

Data are needed for GIS use. These may be provided by the software developer, government agencies, GIS specialists, or developed by teachers. The following sources provide data compatible with most GIS systems.

Our World GIS Education This is a series of geography textbooks that include U.S.-based datasets. The textbook includes a free ArcGIS desktop version for 1 year, with data in the appendix DVD. Four textbooks complete the series: Napoleon and Brook (2008) for primary schools, Palmer et al. (2008a, 2008b) for junior and senior high schools, and Keranen and Kolvoord (2008) for universities.

Itochiri and Geo Link Datasets are available on the websites "Itochiri[18]" and "Geo link,[19]" providing data compatible with MANDARA or Google Earth to senior high school teachers. Teaching materials for geographic and Earth scientific experiments in junior and senior high schools with satellite data (mentioned in Sect. 15.3.2.1) provides population density maps, green coverage maps, heat island maps, and some basemap data such as roads or rivers. These can be used in Chizutaro, Google Earth, and QGIS.

The GIS Forum for Education[20] Datasets for Google Earth are provided in the "Google Earth data Bar[21]" web page. They also operate the web page "Data for class[22]" linked on Google Earth data Bar.

Government Datasets The urban planning department in the Fujisawa City government provides GIS data (e.g., roads, rivers, and aerial photos) for schools in the city (Fujisawa City 2007).

15.3.2.4 Training Videos GIS Operations

Data served over web GIS is the easiest technology for classroom use, given that schools have broad-bandwidth internet connections. Without web GIS, teachers would need to use more complex GIS software with prepared datasets. In the absence of internet access and teacher GIS skills, one potential solution is to demonstrate GIS functions with videos. Examples are already posted on the website, YouTube. For example, "Itochiri channel[23]" and "Google Earth geography[24]"

[18] http://itochiriback.seesaa.net/. Accessed 31 July 2012. (J)

[19] http://itcz.web.fc2.com/geolink.html. Accessed 31 July 2012. (J)

[20] https://sites.google.com/site/egisforum/. Accessed 31 July 2012. (J)

[21] https://sites.google.com/site/egisforum/home/detabar. Accessed 31 July 2012. (J)

[22] https://sites.google.com/site/egisforum/home/data4class. Accessed 31 July 2012. (J)

[23] http://www.youtube.com/user/itochiri001?feature=watch. Accessed 31 July 2012. (J)

[24] http://www.youtube.com/playlist?list=PLE20B0B2857CCA840&feature=view_all. Accessed 31 July 2012. (J)

provide videos on the function and steps of operations in Google Earth and Chizutaro. Another is eDesign,[25] where a group of students majoring in GIS at universities have developed "tour" files that operate within Google Earth.

15.4 In-service Teacher Training for GIS

Teachers can learn GIS independently using practice texts written by GIS specialists. Moreover, they can attend training courses taught by GIS specialists, using a manual and datasets.

15.4.1 Practice Texts

Practice texts include GIS exercises, including analysis and mapping skills. Goto et al. (2004) supplied these exercises for use within MANDARA, which address topics such as mapping convenience store distributions and analyzing heat islands. ArcGIS Desktop exercises were furnished by Sadohara et al. (2005) for freshmen to search suitable housing locations within a given distance from railway stations, convenience stores, parking, and highway interchanges. Ito (2010) developed exercises compatible with Chizutaro, MANDARA, and Google Earth, on topics of agricultural trade, plant distribution, and changes of store distribution.

15.4.2 Training with Manual and Datasets

Itoh (2003) delivered in-service MANDARA training for senior high school teachers in 2002. Teachers came from two prefectural research societies for geography education. There were ten teachers, two teaching assistants, and one lecturer in attendance. The training took place over five consecutive weekday nights in the computer room at a university. Since 2002, the Fujisawa City education and culture center has operated in-service Chizutaro training for primary and junior high school teachers twice annually. The exercises focus on developing a disaster prevention map of the school surroundings. The author serves as a lecturer, and other lecturers are teachers who have previously attended the training.

[25] https://sites.google.com/site/edesignt/. Accessed 31 July 2012. (J)

MLIT provides in-service teacher training to promote the educational use of national geographic information, such as National Land Numerical Information,[26] Digital Japan,[27] and Base Map Information.[28] These have been implemented in four prefectures, using Chizutaro, MANDARA, Google Earth, and QGIS to fit the teaching style, facilities, and budget of each prefecture. The exercises are based on mapping disaster prevention around the school. The author and other GIS specialists participated as lecturers, manual writers, and dataset developers. These manuals are available online at "The GIS training program for primary, junior and high school teachers" on the MLIT website.

A number of lessons were learned from these training workshops. First, a majority of teacher questions, approximately 80 %, were attributable to their lack of computer skills. For example, queries included how to double click, where downloaded files can be saved, and how to unzip files. Second, only 10 % of trained teachers used GIS in the classroom, and they required a tremendous effort to prepare the datasets. Therefore, simpler ways to demonstrate and use GIS are essential.

15.5 Conclusions

This chapter described the role of GIS specialists in supporting GIS use in schools. These specialists assist schools in GIS selection that fits teaching styles, available facilities, and budgets. They also provide training workshops.

The experiences point to some commonalities across regions in the teaching of GIS: (1) Reading GIS maps is a higher priority than editing them; (2) delivering datasets via web GIS is the easiest method, as long as the school internet can support the bandwidth; and (3) GIS training videos are an alternative to demonstrate how to create GIS maps in a classroom without the internet.

References

Diercke C (2010) Diercke international atlas. Westermann Schulbuch, Braunschweig

Fujisawa City Education and Culture Center Social Studies Research Committee (2007) Kyouka-ryouiki kenkyu: Atarashii chiki gakusyuu; shiminsei no ikusei; kyouiku GIS wo riyou shite (The research of subjects and field: new regional studies; the fostering of citizenship using GIS). Fujisawa City, Fujisawa. http://www.city.fujisawa.kanagawa.jp/kyobun-c/page100077.shtml. Accessed 31 July 2012

Fukuda T, Tani K (2003) Koukou chiri kyoiku ni okeru GIS riyou no kanousei (The GIS availability in senior high school geography education). Saitama chiri (Saitama Geogr) 27:17–25

[26] http://nlftp.mlit.go.jp/ksj-e/index.html. Accessed 31 July 2012.

[27] http://portal.cyberjapan.jp/site/mapuse/index.html. Accessed 31 July 2012. (J)

[28] http://www.gsi.go.jp/kiban/ Accessed 31 July 2012. (J)

Goto S, Sakai S, Kato I, Tani K (2004) MANDARA to EXCEL ni yoru shimin no tameno GIS kouza (GIS seminar for citizens with MANDARA and EXCEL). Kokon shoin, Tokyo
Ito T (2010) Itochiri shiki: chiri no jugyou ni GIS (The Itochiri methods: GIS for geography classes). Kokon shoin, Tokyo
Ito T (2012) GIS to chiri kyouiku (GIS and geography education). E J GEO 7(1):49–56
Itoh S (2003) Koutou-gakkou chiri no gensyoku kyouin ni taisuru GIS kousyuukai kaisai no kokoromi (GIS training course for high school geography teachers). Chiri Jyoho System Gakkai Kouen Ronbun-shu (Paper Proc Geogr Inform Syst Assoc) 12:249–254
Itoh S (2004) GIS no kyouiku riyouni taisuru shien (The supports for educational use of GIS). Chiri Geppou (Geogr Monthly Rep) 477:7–9
Keranen K, Kolvoord R (2008) Making spatial decisions using GIS: level 4 (Our World GIS Education). ESRI press, Redlands
Murayama Y (2002) Chiri kyouiku you internet GIS no kaihatsu (Availability of internet GIS for geography education). Tsukuba Daigaku Jimbun Chirigaku Kenkyu (Tsukuba Stud Hum Geogr) 26:25–43
Napoleon E, Brook E (2008) Thinking spatially using GIS: level 1 (Our World GIS Education). ESRI press, Redlands
Oshima H, Ota H, Fukui H (2011) Eisei deta handoku you kyouzai no kaihatsu (Teaching materials development for satellite data interpretation). Chizu (Map J Jpn Cartogr Assoc) 49(2):16–23
Palmer A, Palmer R, Malone L (2008a) Mapping our world using GIS: level 2 (Our World GIS Education). ESRI press, Redlands
Palmer A, Palmer R, Malone L, Voigt C (2008b) Analyzing our world using GIS: level 3 (Our World GIS Education). ESRI press, Redlands
Sadohara S, Kawasaki A, Yoshida S, Furuya T (2005) Zukai! Arc GIS mijikana jirei de manabou (Illustrated ArcGIS! Let's learn it with familiar cases). Kokon shoin, Tokyo
Tani K, Sato T, Ohnishi K, Okamoto K, Okunuki K (2002) Chugakkou ni okeru chiri kyouiku you GIS no kaihatsu to kyouiku jissen (Development of an educational GIS and its use on a trial basis in junior high school classes). GIS Riron to Ouyou (Theor Appl GIS) 10(2):69–77
Ugawa Y, Kamiya H, Itoh S (2003) Digital chizu to sono server no kyouiku riyou ni kansuru ikutsuka no torikumi (Studies of digital maps and their servers for educational use). Chiri Jyoho System Gakkai Kouen Ronbun-shu (Papers Proc Geogr Inform Syst Assoc) 12:211–216

Part IV
Research on Geography Education in Japan and Other Countries

Chapter 16
Trends in Japanese Geography Education Research in Recent Years

Hiroya Yoshimizu

Abstract The aim of this chapter is to examine research trends in Japanese geography education, published in academic journals and books from the 1980s to the present. Geography education refers here to school education, namely, elementary, junior high, and senior high school. Attention is paid to several types of research areas in the country, including cognitive, historical, overseas, and practical research such as curriculum or lesson plan development. Recent research on geography education has been influenced by the new national curriculum in Japan and other countries. However, there is a paucity of work in the area of lesson and evaluation studies, which focuses on curriculum evaluation and instruction lessons developed from educational standpoints. Therefore, increased accumulation of research results is expected in the future.

Keywords Geography curriculum • History of geography education • Japanese geography education research

16.1 1970s and 1980s

Japanese scholars are active in geography education research. This section describes the types of research and their contributions to capacity building and strengthening of the field in Japan.

16.1.1 Cognitive Research

Human geography became prominent and influenced geography education during the 1970s. Experimental geography education research focused on spatial and locational awareness. Researchers aimed to experimentally measure changes in children's experiences, perceptions, and awareness (Iwamoto et al. 1985).

H. Yoshimizu (✉)
Hyogo University of Teacher Education, 942-1 Shimokume, Kato, Hyogo, 673-1494, Japan
e-mail: yosimizu@hyogo-u.ac.jp

Y. Ida et al. (eds.), *Geography Education in Japan*, International Perspectives in Geography 3, DOI 10.1007/978-4-431-54953-6_16

Iwamoto (1981) studied the imaginary environment of children and structure of their local environments while referencing Hart's (1979) theories. Iwamoto successfully measured the spatial domains of children living in cities. This research is a valuable resource, rich in experimental results that are foundational to geography curriculum development and the selection of appropriate instructional content. Other Japanese researchers continued this line of research, for example, Teramoto (1984) analyzed hand-drawn maps to understand the activity domains of children doing fieldwork. Ohnishi (1999) extended Teramoto's work to investigate children's understanding of the environment.

16.1.2 Research on History of the Theory of Geography Education

Research on the history of geography education covers a range of topics, such as on the underpinning theories of geography education and historical practices of geography theory in education.

Taro Tsujimura and Kumataro Yamamoto, two geographers in the early twentieth century, contributed to the idea of Landscape Theory in Japan, a concept that is part of geography education. Thus, observation was considered an important means for collecting information and learning about surrounding areas. Gaps in their work were pointed out by Iwata (1985a), who criticized Tsujimura for not including the core idea of Schulter's Landscape Theory. Iwata (1985b) analyzed practices of geography education using Yamamoto's philosophy that combines the natural and cultural landscape, grounded in the Landscape Theory. Iwata noted the contribution of Yamamoto's practices, because they used quantitative analysis that differed from his philosophy.

16.1.3 Research on Geography Lesson Theory

Iwata (1986) argued that teachers should use geography textbooks in conjunction with clear lectures developed on the components of knowledge and geographical perspectives and way of thinking. He presented a lecture structure that allows students to search for concepts in geography, enabling them to differentiate between "understanding" and "knowing." His work was based on social science research studies. Given the recent influence of the Programme for International Student Assessment (PISA) test, there is a trend toward promoting explanations and descriptions. However, this type of explanation-based learning was already suggested by scholars (e.g., Iwata 1991). Iwata was a visionary for geography, with progressive ideas. For example, he saw that Social Studies could help students develop good citizenship and that rational decision making was at its foundation. He conducted a value analysis and proposed Social Studies classes as a way for students to develop general inquiry and analysis values. Explanation and decision-making learning techniques were agreeable to teachers and fit course content. Thus, these proposed teaching styles became an impetus to the development of new geography education content and promoted further research on Social Studies education.

16.1.4 Geographical Simulation Teaching Materials

In the 1970s teaching community, there was a shift from a real knowledge towards a constructionist approach to building knowledge. This shift led to a decrease in traditional regional geography studies and an increase of simulation teaching materials developed in England. In Japan, these materials attracted attention from the education community, which saw entertainment as a learning method. Simulation teaching was introduced in the school systems in the 1980s, and a report on international implementation was published. For example, Watanabe (1984) introduced a simulation game in geography textbooks in England and Nishiwaki (1989) reported the practice with original simulation resources. Research on simulation teaching and materials in Japan was treated in a book (Yamaguchi et al. 1993).

16.2 The 1990s

In the 1990s, a geography curriculum analysis of the U.S. revealed notable approaches, which spawned interest in the field of geography education.

16.2.1 Curriculum Analysis

The National Curriculum Standards in Japan was created after World War II, under the supervision of the General Headquarters (GHQ). The provisional guidelines were initially created in 1947 and since, they have been periodically revised and serve as a legally binding document. Therefore, the motivation for elementary and junior high school teachers to individually develop their own curricula did not arise. Schools only needed to teach according to authorized textbooks. A byproduct of this education system was that teachers lost interest in the mandated curriculum. However, Ito (1994) argued that the greatest mistake and crime of Social Studies is that history and geography were included within the same subject. The delicate balance between geography and history within the mandated Social Studies class has sparked countless debates. Social Studies and geography education researchers studied the curriculum within the courses offered and focused on foreign curriculum theories, in hopes of gaining insight from foreign practices.

Nakayama (1991) reported on the geography education movement in the U.S. from 1987 to 1988. In particular, the state of Minnesota's "Elementary and Junior

High School Geography Education Guidelines" were translated into Japanese as a model. Nakayama's research received strong interest because it reported efforts in the U.S. to reorganize geography education, at a time when Social Studies in senior high school was being deconstructed in Japan. Ando (1993) studied the Social Studies curriculum in the U.S. and proposed a curriculum theory that extended Hanna's (1956) theory "expanding communities of men," from an article entitled "Society-Child-Curriculum". Hanna's theory has since been modified and incorporated into Japan's Social Studies curriculum. The expanding environments theory developed by Ando is a foundation of Japan's elementary school Social Studies curriculum. Another model U.S. project is the High School Geography Project (HSGP). Its curriculum structure, although unpopular among U.S. geographers and geography education researchers, had a well-balanced design for teaching Social Studies in Japan. Kusahara's (1996) study was groundbreaking and changed attitudes toward the HSGP curriculum structure. That study serves as a foundational reference when discussing geography education within the subject of Social Studies.

16.2.2 International Understanding

Knowledge about international countries has been incorporated in the study of foreign countries at the high school level since the 1970s. The aims of such study are to promote better understanding of other countries and to teach students about foreign lifestyles and cultures. For example, Nishiwaki (1993) discussed the aim of developing world citizenship through geography education. He argued that education about international issues should not be just simple memorization or even understanding facts, but that it should promote student abilities to participate in a global society. Nishioka (1996) was the first geography education specialist to write a systematic development studies book. His aim was to understand causality between poverty in developing countries and class differences between the northern and southern hemispheres. He also discussed how people can change their attitudes to overcome these issues.

16.2.3 Geographical Perspectives and Way of Thinking

The concept of "geographical perspectives and way of thinking" has its roots in the mid 1950s. This concept holds that there are conceptual theories and academic skills unique to the subject of geography. This perspectives and way of thinking is the academic skill set that many geography education theorists believe should be developed in students, and its official inclusion into education was announced in 1969. However, the meaning of this concept is not unanimously applied as a skill set to student learning in high school geography guidelines (Toida 1999).

16.3 The 2000s

In the 2000s, a curriculum was developed to support geographic decision-making and social participation, including Education for Sustainable Development (ESD) as contextual background. Regional geography education is emphasized in the National Curriculum Standards that was introduced in 2008/2009. This section summarizes influential Japanese research that has shaped geography education.

16.3.1 Curriculum Development

Yamaguchi (2002) published novel research findings on student development of geography awareness. He believed that children's geography awareness progresses in three stages: (1) A preparation period, (2) development period, and (3) deepening period. The latter period is also known as the "explosive growth period," which is crucial to understand in order to create a successful curriculum for geography education. Yamaguchi's views are persuasive because of the large amount of data presented.

Saito (2003) was influenced by Piaget's research on childhood cognitive development, and proposed the developmental theory of geography education. This theory strives to match "spatial experiences," "knowledge acquisition" of the world, and "skill acquisition" with a child's developmental progression as a foundation. The aim is to teach international understanding, international politics, and environmental problems for connecting the child with modern society. This perspective has led to new curriculum theories and experimental research on children's spatial cognition.

Kusahara (2004) believed that geography education is a purposeful tool and central subject that can develop the next generation of leaders in the nation and society. He developed a new framework described as education to develop leaders of a democratic nation and society, to rethink theories governing geography education. He borrowed geography education research in the U.S. as reference in framework formation. Kusahara concluded that the contribution of geography education to the development of its citizens is desirable for the development of scientific social awareness. His research is broadly accepted and found in curricula and book chapters, and he is recommended as a speaker in lecture schedules. His work also communicates the importance of geography in Social Studies education.

Yoshimizu's (2002a, b) research focus was on learning content in Social Studies. His work suggests that the structure of teaching materials would be effective if organized around ideas of location and distribution (e.g., industry locations in Japan).

16.3.2 Geography Education History

The history of geography education and its structure has been investigated by some researchers over time. Gion (2009) examined the history of geography education in Fukuoka Prefecture, before and after World War II. This work clarified that Social

Studies education during the Meiji and Taisho periods contained early forms of a developing social scientific awareness.

Nagata (2006) studied liberal education reform in Social Studies during the Taisho era. He investigated "lifestyle studies" taught in Asakusa elementary schools in the early Showa period, and analyzed the geography curriculum during this period. He described the history of national education curricula and showed their contrasting structure to world/country spatial formations, from the perspective of "production of space" by H. Lefebvre (Nagata 2009).

16.3.3 Overseas Research

Japanese researchers look to international methods of promoting geography learning. Murayama (2005) discussed the Swedish textbook "The Wonderful Adventures of Nils," a novel used in upper-level Swedish elementary schools. She used the textbook to provide insight into the joy of learning geography from a child's perspective. From there, rich geographic contents were spun into the story, for example, the importance of using different scales to learn about phenomena. Through this literature, geography education promotes awareness and respect of regional diversity and love for the Earth as the supporter of life. Murayama suggested that this may be what is lacking in Japan's geography education.

Tabe (2008) focused on textbooks written by Morse and Goodrich, both literary figures. The former is recognized as the "father of American geography" and the latter known for Parley's Universal History. Tabe explained the process and current structure of modern geography education in the U.S. and Europe. Other researchers have published about western countries, with Ida (1995) reporting in detail about the popularity of geography education in New Zealand. Arai (2011) discussed the content of citizenship development education in junior high schools in that country.

Shimura (2010) described the state of geography education in England through *Geography: The National Curriculum for England*. This study provided a big picture of the English education system by interviewing researchers there and visiting local elementary and secondary schools. What emerged were perspectives on strong topics in environmental geography, including sustainable changes, studies on textbooks, teaching materials (e.g., maps) and work schemes, geography education revival in the country, graphicacy, sample studies, and case studies. This research was comprehensive and detailed. It is thus a valuable reference, since geography education in England is frequently used as a model in Japanese discussions.

16.3.4 Geographic Information Systems (GIS)

The role of GIS in Japan's schools is important, although obstacles are slowing its adoption in schools. Murayama (2004) compiled GIS theories related to geography research, and Ito (2010) developed a low-budget GIS program that can be easily used in the classroom. For a more thorough discussion on GIS education in Japan, please read Chap. 14.

16.4 Geography Education After 2011

In recent years, researchers have suggested curricula and lesson development to propel geographical decision-making and social participation as a background to Geography education as ESD. Studies on regional geography have been conducted because they were valued in the 2008 and 2009 National Curriculum Standards. In addition, the topic of citizenship construction is increasingly represented in those curricula.

Ito (2012) reported a trend in modern geography education in the UK, with comparison to the theoretical claims and curriculum of Standish and Lambert. That article is valuable because it details the national curriculum revision trend in the UK, in conjunction with their outlook on knowledge and regional geography study.

Nakayama et al. (2011) introduced concrete examples for incorporating the perspective of the UN report entitled "Decade of ESD". They also suggested examples for the development of geography education. In addition, Izumi et al. (2012) suggested making "social participation" a keyword in ESD. Ohnishi (2008) translated the Lucerne Declaration, which is a valuable resource that touches on the sustainable development of geography education

Yoshimizu (2011) used the concept of geographic scale to design curricula with a multi-scale approach. Using this as a model, Takeuchi (2012) proposed geography education based on various scales that includes a multi-layered citizenship resource.

16.5 Conclusion

A review of Japanese research on geography education demonstrates the broad and detailed nature of this field. It also indicates a scarcity of research on topics such as geographic theories, student knowledge acquisition, and the use of teaching evaluations to improve geography learning. These are some of the topics in need of further research. To build cumulative capacity, geography education research should collaborate with other fields such as philosophy, educational studies, and cultural humanities among other subjects.

References

Ando T (1993) Doushin-en kakudai ron no seiritsu to hihanteki tenkai: America shougakkou curriculum kousei genri no kenkyu (Establishment of expanding concentric circles theory and critical developments: study of elementary school social studies curriculum composition in the United States). Kazamashobo, Tokyo

Arai M (2011) New Zealand no chuutoubu shakaika gakusyu to koutoubu chiri gakusyu: Shiminsei wo juushi suru shakaika chiri kyouiku (Social studies and geography in secondary education in New Zealand: geography in social studies that hold training citizenship in great account). Shin Chiri (New Geogr) 59(1):6–27

Gion Z (2009) Shakai ninshiki keisei no shiten kara mita Fukuoka ken chiri kyouiku jissenshi (History of geography lessons in Fukuoka Prefecture from the viewpoint of social recognition formation). Kaichosha, Fukuoka

Hanna PR (1956) Society-child curriculum. In: Hunnicutt CW (cd) Education 2000 AD: perspective in elementary education. Syracuse University Press, New York, pp 165–199
Hart R (1979) Children's experience of place. Irvington Publishers, New York
Ida Y (1995) New Zealand no chiri kyouiku (Geographical education in New Zealand: the factors of popular subject). Shin Chiri (New Geogr) 43(1):1–14
Ito R (1994) Shakaika kyouiku no kouzai (Merits and demerits of social studies education). Shakaika kyoiku ronso (Bull Jpn Social Stud Res Assoc) 41:26–33
Ito T (2010) Ito chiri shiki chiri no jugyo ni GIS (The Itochiri methods: GIS for geography classes). Kokon shoin, Tokyo
Ito N (2012) Igirisu ni okeru chiri curriculum ronsou: Standish to Lambert no kyouiku ron ni chuumoku shite (A study of the debate about school geography curriculum in England: with a focus on the Standish vs. Lambert Controversy). Shakaika kenkyu (J Edu Res Social Stud) 76:11–20
Iwamoto H (1981) Kodomo no shinzo keikan ni okeru mijikana chiiki no kouzou (The imaginary environment of "community" in a children's mind). Chirigaku hyoron (Geogr Rev Jpn) 54:127–141
Iwamoto H, Ando M, Teramoto K, Yoshida K, Matsui M (1985) Kodomo no shinriteki hattatsu ni kansuru chirigakuteki kenkyu (Recent trend of geographic approaches to the development of children's environmental perception, cognition and image). Shin Chiri (New Geogr) 33(2):28–38
Iwata K (1985a) Tsujimura Taro no keikan chirigaku to shisouteki gensen (Landschaftskunde developed by Taro Tsujimura and sources of his ideas). Shakaika kenkyu (J Edu Res Social Stud) 33:1–14
Iwata K (1985b) Yamamoto Kumataro no keikan chiri kyouikuron to shisouteki gensen (Teaching of Landshaftskunde developed by Kumataro Yamamoto and sources of his ideas). Hyogo kyoiku daigaku kenkyu kiyo (Hyogo Univ Teacher Edu J) 6:17–131
Iwata K (ed) (1986) Chiri kyoukasho wo riyou shita wakaru jugyou no kaihatsu (Development of clear lectures using the geography textbook). Meijitosho shuppan, Tokyo
Iwata K (1991) Shougakkou shakaika no jugyou bunseki (Analysis of elementary school social studies classes). Tokyoshoseki, Tokyo
Izumi T, Umemura M, Fukushima Y, Ikeshita M (eds) (2012) Shakai sankaku no jugyou dukuri: Jizoku kanou na shakai ni mukete (Development of the lesson plans for social participation: for building a sustainable society). Kokon shoin, Tokyo
Kusahara K (1996) Shakai kagakuka toshite no chiri kyouiku: HSGP no saihyouka (A theory on the teaching geography as social science education: re-evaluation on "the high school geography project"). Shakaika kenkyu (J Edu Res Social Stud) 44:41–50
Kusahara K (2004) Chiri kyouiku naiyou henseiron kenkyu: Shakaika chiri no seiritsu konkyo (Geography education content theory research: reasoning for the establishment of geography in social studies). Kazamashobo, Tokyo
Murayama Y (ed) (2004) Kyouiku GIS no riron to jissen (Theory and application of educational uses of GIS). Kokon shoin, Tokyo
Murayama T (2005) Nils ni manabu chiri kyouiku (Learning geography education from nils: the origin of Sweden's environmental society). Nakanishiyashoten, Kyoto
Nagata T (2006) Taisho jiyu kyouikuki ni okeru shakaikei kyouka jugyou kaikaku no kenkyu: Shotou kyouiku dankai wo chushin ni (Study on the social studies lesson reform in the new teaching in Taisho Era). Kazamashobo, Tokyo
Nagata T (2009) Sekai kokudo kuukan hensei ni taiji suru kyoudo kyouiku curriculum: Tokyo shi Asakusa jinjo shougakkou no seikatsuka jissen no baai (Community education curriculum confronting the world and national spatial configuration: a case study of life-environment studies in Asakusa elementary school in Tokyo City). Shakaika kenkyu (J Edu Res Social Stud) 71:11–20
Nakayama S (1991) Chiri ni mezameta America: Zenbei chiri kyouiku fukkou undou (America's awakening towards geography: America's movements for the revival of geography research). Kokon shoin, Tokyo
Nakayama S, Wada F, Yuasa S (eds) (2011) Jizoku kanou na shakai to chiri kyouiku jissen (Sustainable society and applications to geography research). Kokon shoin, Tokyo

Nishioka N (1996) Kaihatsu kyouiku no susume- Nanboku kyousei jidaino kokusai rikai kyouiku. (Development education: international understanding education of the South and North coexistence era). Kamogawashuppan, Kyoto

Nishiwaki Y (1989) Simulation kyouzai ni yoru toshi no hattatsu no tenkai: Chiba shi wo rei to shite (Lesson plan of "the development of the city": a case of Chiba City). Shin Chiri (New Geogr) 37(2):41–49

Nishiwaki Y (1993) Chiri kyouikuron josetsu (Introduction to geography education theory). Ninomiyashoten, Tokyo

Ohnishi K (1999) Tegaki chizu kara mita kodomo no chikaku kankyou: Sanson no jirei (Children's environmental perception from sketch map: a case study on mountain village in Japan). Shin Chiri (New Geogr) 47(1):1–13

Ohnishi K (2008) Jizoku kanou na kaihatsu no tameno chiri kyouiku ni kansuru Lucerne sengen (Translation of declaration for sustainable development of geography education). Shin Chiri (New Geogr) 55(3/4):33–38

Saito T (2003) Hasseiteki chiri kyouikuron: Piaget riron no chiri kyouikuronteki tenkai (Developmental geography education theory: expanding Piaget's theory to geography education theory). Kokon shoin, Tokyo

Shimura T (2010) Gendai Igirisu chiri kyouiku no tenkai (Recent development in geography education in England). Kazamashobo, Tokyo

Tabe T (2008) America chiri kyouiku seiritsushi kenkyu (Research on the establishment of geography education in America). Kazamashobo, Tokyo

Takeuchi H (2012) Chiiki ni okeru shakai sanka to chiri kyoiuku (Social participation in the community and geographical education). E J GEO 7(1):65–73

Teramoto K (1984) Kodomo no chikaku kankyo no hattatsu ni kansuru kisoteki kenkyu: Kumamoto ken Asodani no baai (A study on the development of children's perception of the environment). Chirigaku hyoron (Geogr Rev Jpn) 57(Ser.A)(2):89–109

Toida K (1999) Chiriteki mikata kangaekata no kisoteki kousatsu (Basic discussions on geographical perspectives and way of thinking). In: Inoue S, Aizawa Y, Toida K (eds) Atarashii jugyo no susumekata: Mikata, kangaekata wo sodateru (A new way to teach geography: foster geographical perspectives and way of thinking). Kokon shoin, Tokyo, pp 8–23

Watanabe A (1984) Igirisu no chiri kyoukasho ni miru simulation game (Simulation games in geography teaching in England). Shin Chiri (New Geogr) 32(1):1–12

Yamaguchi Y (2002) Shakaika chiri kyouikuron (Social studies and geography education theory). Kokon shoin, Tokyo

Yamaguchi Y, Nishiwaki Y, Umemura M (1993) Simulation kyozai no kaihatsu to jissen: Atarashii chiri gakushu no kokoromi (Development and practice of the simulation teaching materials: new trial of the geography learning). Kokon shoin, Tokyo

Yoshimizu H (2002a) Ichi to bunpu gainen ni kansuru mondai hakken kouzou: Nichi-ei kyoukasho bunseki wo to-oshite (Problem-finding structure of "location and distribution" in geographical education: from the analysis of the textbooks used in Japan and UK). Shin Chiri (New Geogr) 49(4):18–31

Yoshimizu H (2002b) Mondai hakken nouryoku wo ikusei suru chuugakkou shakaika chiri jugyou no sekkei: Tangen Nippon no kougyou chiiki no kaihatsu (The design of geography lesson in social studies to develop the problem-finding ability: development of teaching materials for the unit "industrial area in Japan"). Shakaika kenkyu (J Edu Res Social Stud) 57:61–70

Yoshimizu H (2011) Chiriteki scale no gainen wo mochiita multi-scale chiri jugyou no kaihatsu: Chuugakkou shakaika chiriteki bunya mijikana chiiki no chousa Kouchi ken Haruno chiku wo daizai ni (Development of multi-scale geography lessons based on the concept of geographical scale: the unit "survey of home regional area at Haruno Town, Kochi City" in junior high school). Shin Chiri (New Geogr) 59(1):1–15

Chapter 17
Trends of School Geography in Asia

Hyunjin Kim

Abstract To understand current trends of school geography in Asia, this chapter compares and contrasts Geography curricula in Singapore, Hong Kong, South Korea, and Japan. It also identifies similarities and differences between knowledge, skills, and values in the junior high school Geography curriculum. The current curriculum in these countries is designed by merging several approaches, and promotes inquiry-based learning for students to develop a wide range of skills. Furthermore, school geography provides students with an understanding of human and environmental relationships, as an especially important subject in education for sustainable development.

Keywords Asia • Current trends • Junior high school curriculum • School Geography

17.1 Introduction

Historically in Asia, school geography is offered to students in two ways, either as a separate subject or as a part of an integrated subject (Lam et al. 2006). Geography in Singapore and Hong Kong are strongly influenced by British tradition. South Korea and Japan, however, follow the U.S. and geography is embedded in social studies. These countries have undergone curriculum reform in the past 10 years. Although the geography curriculum contains objectives and contents that relate to regional and national needs, there is also a need to address globalization.

This chapter is presented in three parts. It begins with a brief description of geography's place in the school curriculum. Next, it analyzes curriculum structure, focusing on the junior high school level. In the third part, the importance of geography knowledge, skills, and values in the junior high school curriculum is described.

H. Kim (✉)
Hokkaido University of Education, 9 Hokumoncho, Asahikawa, Hokkaido 070-8621, Japan
e-mail: kim.hyunjin@a.hokkyodai.ac.jp

Y. Ida et al. (eds.), *Geography Education in Japan*, International Perspectives in Geography 3, DOI 10.1007/978-4-431-54953-6_17

17.2 Placement of Geography in the School Curriculum

Geography is treated differently among school levels in Singapore, Hong Kong, South Korea, and Japan (Table 17.1). In the elementary school curriculum, geography does not exist as a separate subject. Instead, geographic contents are taught through integrated subjects that are compulsory, for example, in General Studies in Hong Kong or Social Studies in Singapore, South Korea, and Japan. Such integrated subjects in elementary school are based on the expanding environments approach, so geographic content is very prominent within the subjects.

Nevertheless, a recent trend is for integrated subjects to replace geography, which has traditionally been a separate subject in junior high school in many countries. In senior high school, Geography remains a separate subject, but an elective one.

At junior high schools in Singapore, geography is a compulsory subject for the Special, Express and Normal (academic) Courses, but Social Studies replaces it in the Normal (technical) Course. In senior high school, Combined Humanities is a compulsory subject, whereas geography is elective for the Special, Express and Normal courses. In Hong Kong, geography is a separate subject in junior and senior high school, but an elective subject within the Key Learning Area of Personal, Social and Humanities Education (PSHE). As with Social Studies, life and society, and integrated humanities, some integrated subjects in the PSHE include geographical contents.

In South Korea, geography is one component of "Intelligent Life" in first and second grades and of Social Studies from third through tenth grades. In eleventh through twelfth grades, it is taught through two separate subjects, Korean Geography and World Geography. Like South Korea, Japan adopts Life Environmental Studies in first and second grades and Social Studies from third through ninth grades. In Social Studies, geography is one component in elementary school, but a separate subject in junior high school. In senior high school, World History is a compulsory subject; Geography and Japanese History are elective subjects in Geography and History.

Table 17.1 Placement of Geography in the School Curriculum

	Singapore	Hong Kong	South Korea	Japan
Elementary School	Integrated	Integrated	Integrated	Integrated
	Compulsory	Compulsory	Compulsory	Compulsory
Junior High School	Integrated/Separate	Integrated/Separate	Integrated	Separate
	Compulsory	Elective	Compulsory	Compulsory
Senior High School	Integrated/Separate	Integrated/Separate	Integrated/Separate	Separate
	Compulsory/Elective	Elective	Elective	Elective

17.3 Structure of Junior High School Curriculum for Geography

The Geography curriculum is commonly structured via two main approaches, regional or thematic. The thematic approach is further classified into systematic, issues-based, and systems (IGU-CGE 1992). Some western countries made curriculum compromises by merging these approaches (Haubrich 2006). Recently, this trend has been found in Asia.

17.3.1 Systematic and Issues-Based Approaches in Singapore

The junior high school geography curriculum was implemented in 2006. Table 17.2 summarizes curriculum contents. There are a total of five themes, three to be covered in the seventh grade and two in the eighth grade. The curriculum adopts the systematic approach "to show how relationships between people and environment have given rise to the distinctive character of places" (MOE 2005, p. 2). Hence, the themes primarily include scientific content related to physical and human geography. Case studies are used to explain physical-human relationships in some regions. For example, in the case study "Living with Volcanoes" in Indonesia, pupils evaluate benefits and risks of living in volcanic areas (MOE 2005). The curriculum also supports the issues-based learning approach, as "current issues and events should be incorporated into the lessons to ensure that the subject remains relevant and interesting" (MOE 2005, p. 2).

Table 17.2 Contents of Junior High School Curriculum for Geography in Singapore

Seventh Grade	Theme I: Introduction to Geography	1 Overview of Geography
		2 Earth as Home—Solar System
	Theme II: Understanding the Environment	3 Physical and Human Environments
		4 Environments Through Maps
		5 Environments Through Photographs
	Theme III: The Physical Environment	6 Landforms and Rocks
		7 Rivers
		8 Weather and Climate
		9 Natural Vegetation
Eighth Grade	Theme IV: The Human Environment	10 Population and Settlements
		11 Agriculture
		12 Transport and Communications
	Theme V: Managing the Changing Environment	13 Land Supply
		14 Water Resources
		15 Pollution
		16 Global Warming and Ozone Depletion

Source: MOE 2005

17.3.2 Issues-Based and Regional Approaches in Hong Kong

In 2011, the new junior high school curriculum for geography was completed. Table 17.3 shows that the curriculum consists of three sections, with twelve modules (four modules in each section). Students study nine modules (six core and three elective modules) from seventh through ninth grades. Each module focuses on issues at scales from local to global. For example, Section A, "From Hong Kong to the World," is designed to explore issues (e.g., urban environment, natural hazards, tourism, and changing climate) in a local context, and then extends to the national, regional, and finally global scale. Issues-based and regional approaches "aim to provide students with a contemporary world view and foster their ability to look at regions with the perspective of change and development" (CDC 2011, p. 14).

17.3.3 Systematic and Issues-Based Approaches in South Korea

Until the curriculum reform launched in 2007, junior high school geography had been focused on Korean and world geography. Under the regional approach, students memorized large quantities of geographical facts. Hence, the new junior high school curriculum changed to the thematic approach, mixing systematic and issues-based approaches. Table 17.4 presents geographic contents in the social studies curriculum. There are 11 units with geographic content, six units to be covered in the seventh grade and five in the ninth grade. The units are designed for students to learn physical and human patterns through regional case studies (MEST 2007).

Table 17.3 Contents of Junior High School Curriculum for Geography in Hong Kong

A: From Hong Kong to the World	Using Urban Space Wisely—Can we maintain a sustainable urban environment?	
	Living with Natural Hazards—Are we better equipped than the others?	
	Tourists—Friends or foes?	Either One
	Changing Climate, Changing Environments	
B. From China to the World	Food Problems—Can we feed ourselves?	
	The Trouble of Water—Too much and too little	
	Population Problems—Just about numbers?	Either One
	Taming the Sand—A long-lasting combat against desertification and sandstorms	
C: Challenges for our world	Global Shift of Manufacturing Industry—Opportunities and threats	
	Scramble for Energy	
	The Geography of Disease—Facing a spreading risk	Either One
	Oceans in Trouble	

Source: CDC 2011

Table 17.4 Geographic Contents of Junior High School Curriculum in South Korea

Seventh Grade	The world we live in
	Various climate regions and life of people
	Various types of landforms and life of people
	Different cultures by region
	Population change and population problems
	Urban growth and urban problems
Ninth Grade	Use and management of resources
	Industrial activity and change of region
	Different environmental problems by region
	My country in the world
	The future of united Korea

Source: MEST 2009

Table 17.5 Contents of Geographical Field in Junior High School Curriculum in Japan

Seventh and eighth Grade	(1) Various regions in the world	A. Regional structure of the world
		B. Different lifestyles and environments in the world
		C. Regions in the world (6 Continents)
		D. Geographical survey of various regions in the world
	(2) Various regions in Japan	A. Regional structure of Japan
		B. Outlook of Japan compared to the world
		C. Japanese regions (7 regions)
		D. Geographical survey of surrounding area

Source: MEXT 2008

For example, in the unit "Various Climate Regions and Life of People," students begin by showing global climate patterns, and then exploring characteristics of the climate in Korea before comparing it with that in other countries.

17.3.4 Topical and Dynamic Regional Approaches in Japan

The Ministry of Education, Culture, Sports, Science and Technology (MEXT) revised the junior high school geography curriculum in 2008. It consists of eight units taught over 2 years (seventh and eighth grades). Table 17.5 shows that the regional approach plays a strong role in the curriculum, which consists of two sections, Japan and the world. A danger of this approach is that there is too much factual knowledge to learn, such as place names. However, adoption of a regional approach in the new curriculum is not a "cape and bay" method. The curriculum requires students to learn through topical and dynamic regional studies taught in the unit "Regions in the World," with particular geographic topics selected to represent continental characteristics.

Dynamic regional studies are taught in the unit "Japanese Regions." Japan is divided into seven regions: Hokkaido, Tohoku, Kanto, Chubu, Kinnki, Chugoku and Shikoku, and Kyushu and Okinawa. These correspond to seven themes in the curriculum: physical environment, historical background, industry, environmental issues/preservation, population and urban/rural settlements, lifestyles/cultures, and connections with other regions. Students should learn one region by focusing on one main theme and related sub-themes.

In summary, the curriculum for geography has shifted from a descriptive, regional approach to a positivist, systematic one and now an issue-based one. Through a regional perspective, geography enables students to understand and construct knowledge about the world. Therefore, criteria for selecting regions in the curriculum may include "major issues in the contemporary world," "geographical perception of space, place and environment," "geographical ways of looking at spatial organization," "illustrative examples," "student experiences, interests and preconceptions," "significance for the individual, people, culture and the environment," and "balance" (IGU-CGE 2007).

17.4 Knowledge, Skills and Values in Geography Education

In this part, three classes of objectives; knowledge, skills and values are discussed in the context of teaching and learning.

17.4.1 Factual and Conceptual Knowledge

In a survey conducted by Gerber (2003), knowledge was regarded as including fact and concept. Concept is essential for understanding fact, so many people have emphasized it in geography education. There are two geography learning trends in each region and country. One is the conceptualization of geography in Singapore and Hong Kong, whereas the acquisition of solid and broad knowledge is emphasized in South Korea and Japan.

As a separate subject in junior high school in Singapore and Hong Kong, geography has a more conceptual curriculum than that in integrated subjects. In Singapore, understanding of humans and their relationship with the environment is one of the aims of the junior high school geography curriculum (MOE 2005). Hence, students learn geographic concepts about human-environment relationships. For example, in Theme II, "Understanding the Environment," students differentiate between physical and human environments, and contrast various human-environment relationships. According to Chang (2011), such conceptual understanding of those relationships has been key in the past 30 years in Singapore education.

Like Singapore, an aim in Hong Kong's junior high school geography curriculum is "to enable students to develop knowledge and understanding of space, place and environment, in particular the spatial arrangement of places and the interaction between humans and the environment" (CDC 2011, p. 9). The curriculum is designed to facilitate student understanding and application of key geographical concepts, including space, place, region, human-environment interaction, global interdependence, and sustainable development (CDC 2011).

In South Korea, junior high school geography is subsumed under Social Studies. Although students acquire some geographic concepts, they are limited. Since learning about concepts requires a deeper understanding beyond facts, it is difficult to provide a comprehensive curriculum of geographic concepts within Social Studies. This difficulty is compounded by the fact that Social Studies teachers may not have majored in geography. Kim and Ryu (2006) argued that the integration of Geography, World History, and Civics in Social Studies textbooks actually narrows student understanding of geography. Conversely, geography in Social Studies can provide students with a spatial perspective of various social problems and issues.

In Japan, the current junior high school geography curriculum emphasizes the acquisition of basic knowledge and concepts, and is a separate subject that is part of Social Studies. Although there is a conceptual curriculum for geography as in South Korea, it is taught by Social Studies teachers who may not have majored in geography.

17.4.2 Geographic Skills and Inquiry-Based Learning

Geographic skills include finding, analyzing, processing, evaluating, presenting, and using geographical information (Haubrich 2006). The junior high school geography curriculum in Singapore stresses the teaching of geographic skills such as atlas and map reading, as well as photo interpretation. Generic skills in data collection, analysis, communication and application of geographic knowledge have also been integrated into the curriculum (MOE 2005). There are a variety of activities (e.g., group discussions, cooperative learning, simulation exercises, role playing, debates, and field studies) that furnish students with a range of meaningful learning experiences (MOE 2005). In particular, field studies can give students skills that they need to connect to the world.

In Hong Kong, skills are divided into three parts, geographical inquiry skills, geographical skills, and competencies. Geographical inquiry skills are abilities to ask geographic questions, to locate and collect information and data relevant to the inquiry from a variety of sources, to organize and present information and data in appropriate formats, and to analyze and interpret information and data for drawing conclusions. Geographical skills include atlas and map reading, photo interpretation, fieldwork methods and IT techniques. Competencies include communication skills, critical thinking skills, and creativity (CDC 2011).

South Korea's Social Studies curriculum states that students should "develop skills collecting, processing and using information and data to understand social phenomena and issues, skill of inquiry, decision making and social participation to resolve various problems in society" (MEST 2009, p. 4). In addition, the curriculum guideline explains that the "process of geography inquiry and problem solving consist of three steps: (1) observation, perception, definition and classification; (2) application, analysis and synthesis, as well as explanation and prediction; value judgment, problem solving and evaluation, and decision making" (MEST 2007, p. 281).

In Japan, the term "Geographical Way of Seeing and Thinking" is frequently used to describe the aim of geography education. This term was first used in the junior high school curriculum revised in 1969 (Kim 2009). The current curriculum has adopted this term as geographic questioning (MEXT 2008). In addition, there are two types of geographic skills in the curriculum, one is the use of geographic information and the other is map use (MEXT 2008). The former includes finding, selecting, processing, and presenting geographic information, whereas the latter includes reading and interpreting different types of atlases, maps, and plans, as well as presenting geographic information though visual representation.

17.4.3 Education for Sustainable Development

Geography is a major subject that contributes to education for sustainable development (ESD). The Lucerne Declaration on Geographical Education for Sustainable Development states the following:

Geography Education can greatly contribute to achieving the goals of the United Nations Decade of Education for Sustainable Development by providing relevant knowledge, skills, values and attitudes crucial for a peaceful coexistence of individuals with nature on this planet (IGU-CGE 2007).

Many countries in Asia recognize the relationship between geography and sustainable development in the school curriculum. According to Chang (2011), Geography and Social Studies in Singapore highlight ESD, but History and Literature show little evidence of it. The junior high school geography curriculum aims to "provide a holistic understanding of physical-human relationships, to develop an informed concern about the quality of the environment and the future of the human habitat; and thereby to enhance students' sense of responsibility for the care of the Earth and its people" (MOE 2005, p. 1). These aims make it clear that geography is important in ESD.

Lee (2006) indicated that Geography and Biology include ESD concepts at the junior high school level in Hong Kong. Kwan (2003) argued that Geography can play a significant role in helping students develop competency in citizenship as they practice and care for the environment. This learning could foster students' roles in environmental stewardship in an era of educational and curriculum reform. This aligns with the aims of the new junior high school geography curriculum, which mentions "informed and responsible citizens who are willing to act for the betterment of their home city, home nation and the world and to contribute to the sustainable development of human societies and the natural environment" (CDC 2011, p. 14).

In South Korea, ESD is covered under ethics, science, technology and home economics, and Social Studies. According to Lee et al. (2007), for a truly integrated understanding of sustainable development, Geography and Social Studies teachers in South Korea should consider environmental, physical, economic, social, and cultural aspects, as well as relationships across these components.

In 2008, Japan's elementary and junior high school curriculum was revised to provide specific measures to instill in students the "living skills" that they need. The current junior high school curriculum refers to the sustainability of nature, economy, and society. Hence, the ESD concept was enhanced within each subject, including Social Studies and Geography. An example of a suitable ESD topic is pollution (Nagata 2009), which has been integrated in Social Studies and Geography since the 1960s and incorporated into the national curriculum in 1969.

17.5 Conclusion

This chapter presented current trends in school geography across four countries in Asia, Singapore, Hong Kong, South Korea, and Japan. Trends were found via the following case studies.

- The geography curriculum was designed by combining several teaching approaches. Although a descriptive regional approach is not central to this design, region is still a key concept in that curriculum.
- The geography curriculum promotes inquiry-based learning to induce skill development. Through the inquiry process, students develop a wide range of geographical and general skills.
- School geography is an especially important subject for ESD. Geography supports student understanding of human and environmental relationships, an important foundation for their contributions to a sustainable society and environment.

References

CDC (2011) Personal, social and humanities education key learning area geography (secondary 1–3). Curriculum Development Council, Hong Kong

Chang CH (2011) Is Singapore's school geography relevant to our changing world? Rev Int Geogr Edu Online 1(2):141–157

Gerber R (2003) The global scene for geographical education. In: Gerber R (ed) International handbook on geographical education. Kluwer Academic Publishers, Dordrecht

Haubrich H (2006) Changing philosophies in geographical education from the 1970s to 2005; an international perspective. In: Lidstone J, Williams M (eds) Geographical education in a changing world; past experience, current trends and future challenges. Springer, Dordrecht, pp 87–92

IGU-CGE (1992) International charter on geographical education. http://www.igu-cge.org/charter-translations/1.%20English.pdf. Accessed 6 Dec 2013

IGU-CGE (2007) Lucerne declaration on geographical education for sustainable development. http://igu-cge.tamu.edu/LucerneDeclaration.pdf. Accessed 6 Dec 2013

Kim HJ (2009) Geographical inquiry-based learning in Japan. In: Introducing "spatial thinking" to geography education, Proceedings of the IGU-CGE Tsukuba conference (Tsukuba), pp 221–227
Kim HH, Ryu JM (2006) The struggle to revive geography in the national curriculum: identity politics of social studies education in Korea. In: Changes in geographical education: past, present and future, Proceedings of the IGU-CGE symposium (Brisbane), pp 249–253
Kwan T (2003) Geography and citizenship education in Hong Kong. Int Res Geogr Environ Edu 12(1):64–71
Lam CC et al (2006) Geographical education in east Asia. In: Lidstone J, Williams M (eds) Geographical education in a changing world; past experience, current trends and future challenges. Springer, Dordrecht, pp. 139–154
Lee TN (2006) Education for sustainable development–some observations from Hong Kong. In: Changes in geographical education: past, present and future, Proceedings of the IGU-CGE symposium (Brisbane), pp 286–290
Lee JW et al (2007) The status of education for sustainable development in the schools: United States and South Korea. In: Geographical views on education for sustainable development, Proceedings of the Lucerne-symposium (Lucerne), pp 171–177
MEST (2007) Junghakkyo Kyoyukkwajeong Haeseolseo (Guideline of the school curriculum for lower secondary school: social studies). Ministry of Education, Science and Technology, Seoul
MEST (2009) Junghakkyo Sakaigwa Kyoyukkwajeong (The school curriculum for lower secondary school: social studies). Ministry of Education, Science and Technology, Seoul
MEXT (2008) Chugakko gakushu shido yoryo kaisetsu shakai hen (Guideline of the course of study for lower secondary school: social studies). Nihon Bunkyo shuppan, Osaka
MOE (2005) Geography syllabus lower secondary. Ministry of Education, Singapore
Nagata S (2009) Pollution education in social studies as a part of ESD in elementary Schools. In: Introducing "spatial thinking" to geography education, Proceedings of the IGU-CGE Tsukuba Conference (Tsukuba), pp 110–114

Part V
Geography Education in the Future

Chapter 18
Geography Education as Part of Social Studies Education in Japan

Tomoko Murayama

Abstract This chapter discusses the future prospects of geography education in Japan by first examining its current status and associated issues within Social Studies education. The structure of Social Studies is divided into multiple subjects areas (i.e. Geography and History, and Civics) with increasing grade levels. Whereas geographic content forms a large part of elementary school Social Studies, it is taught alongside History and Civics in junior school Social Studies. With the 1989 revision of the National Curriculum Standards, Social Studies in senior high school was reorganized into two subjects "Geography and History and Civics", as well as "World History" that became compulsory subjects. Since this amalgamation, the status of Geography in senior high school has rapidly declined. Geography taught within Social Studies instruction plays a crucial role in developing students' social awareness, a premise of value judgments and decision making for thinking about how to make society better. Geography studies specific phenomena and the relationship between human and nature, making it a valuable contribution to ESD.

Keywords ESD • Geography education • History education • Regional geography • Social Studies

18.1 Introduction

"Geography", in the strict sense, has not existed as a single subject in Japan since the emergence of Social Studies after World War II. The structure of Social Studies changes from the integrated to the subdivided with increasing grade levels. Such that it is broken into different study areas and taught by subject area in junior high school and senior high school, respectively.

Because geography is not taught as an independent subject, any discussion of the ideal state of geography education must consider other disciplines within which it is taught. This chapter aims to shed light on the present situation and issues of Social

T. Murayama (✉)
Ibaraki University, 2-1-1, Bunkyo, Mito, Ibaraki 310-8512, Japan
e-mail: mura@mx.ibaraki.ac.jp

Y. Ida et al. (eds.), *Geography Education in Japan*, International Perspectives in Geography 3, DOI 10.1007/978-4-431-54953-6_18

Studies in a broad sense. This includes geography taught as part of Social Studies in elementary and junior high schools, as well as geography taught within the Geography and History subject in senior high schools. Finally, a discussion of the future role and ideal state of geography education is presented.

18.2 Teachers' and Students' Attitudes Towards Geography

18.2.1 Social Studies in Elementary and Junior High School

According to a survey targeting elementary-school teachers (Benesse Educational Research & Development Institute 2010), the proportion of those who expressed "confidence" in their teaching (i.e., those who rated themselves as "confident" or "somewhat confident") was highest in mathematics, whereas the percentage in Social Studies was very low. The proportion of those who answered "unconfident or somewhat unconfident" for Social Studies was almost half. Even if the number of teaching years is taken into account, the proportion of teachers who had confidence in their teaching of Social Studies never rose above 50.0 % (Table 18.1). These data suggest that the quality of Social Studies classes is dependent upon the teacher.

Unlike elementary schools where teachers are responsible for instruction in all subjects, junior high school teachers instruct specific subjects. Social Studies teachers have varying backgrounds in the arts and humanities, including literature, law, and economics.

A similar survey as above was conducted for junior high school teachers throughout Japan (Iwamoto et al. 2010). The national survey revealed that, of the three subjects of Social Studies, namely, Geography, History, and Civics, teachers had the most confidence teaching History (43.2 %), followed by Civics (29.3 %). Only 14.9 % of teachers had confidence teaching Geography, however, it received the highest proportion of "lack confidence" (30.1 %).

The data makes sense when compared against respondents' university major. The largest proportion of respondents had majored in History (25.1 %); only 11.8 % had studied Geography. Teachers did not lack confidence in their teaching of History, irrespective of their major, but they questioned their ability in Geography. Together, these surveys suggest that teachers' lack of confidence in teaching certain

Table 18.1 Results of survey—percentage of "confident" in teacher responses

Years in teaching	Mathematics	Japanese	Social Studies
Less than 5 years	60.7	32.3	29.0
6–10 years	77.3	34.4	43.1
11–20 years	82.9	48.5	43.6
21–30 years	87.8	62.6	46.9
More than 31 years	92.5	76.7	44.8

Source: Benesse Educational Research & Development Institute (2010)

topics stems from their training experience and evolution of course content and instruction methods. For example, Geography focuses largely on current social phenomena, which is constantly changing. On the other hand, Civics also handles current phenomena as case studies, but the basic knowledge is relatively clear-cut and constant. This contrasts with Geography where there is no clear standards to guide what constitutes basic knowledge and the extent to which it should be taught. Moreover, the teaching content and methods often change with each revision to the National Curriculum Standards. The 1997 revision is viewed as having placed a considerable burden on teachers because it shifted the emphasis from knowledge and comprehension to skills and methods. It was no longer possible to teach from the contents of a textbook. Since many of the younger teachers had not studied Geography in when they were students in senior high school, as it was an elective rather than a mandatory subject then, major instructional differences were observed between those who had and those who had not taken secondary geography.

Another element of geography education is students' perception of geography. A cohort of junior high school students were surveyed about their interest in learning Geography (Arai 2010). More than 80 % of first grade junior high school students responded that they were interested in the subject, particularly in the lives of people around the world. Their desire was to take a fresh look at Japan from a wider perspective, so as to "become friends with people around the world and to make the world a better place. The same student cohort was surveyed again a few years later and the findings showed a rise in the students' interest in people around the world. Evidence of such global-minded attitudes is captured in students' quotes such as "This [subject] needs to be known" and "because we all live on the same planet." The results indicate students' awareness of broadening their perspective and considering world issues in relation to Japan. The quotes "to understand the similarities and differences with Japan," "to review my own activities and way of living," and "to understand what needs to be done" are further examples of an international mindset. These answers point to students' budding desire for global citizenship.

Although many teachers report a lack of confidence in teaching Geography, students have an interest in Geography. Arai (2010) stated that these healthy desires of students must be me". However, something must be done to address teachers' lack of confidence.

18.2.2 Senior High School Geography and History

Geography and History teachers in senior high school give instruction in specific subjects—Geography, Japanese History, and World History. Most teachers majored in their instructional topics in university.

In 2005, a nationwide survey was conducted on attitudes towards courses, targeting senior high school students and their teachers (National Institute for Educational Policy Research 2007). This survey examined the current teaching and learning landscape across subjects under the National Curriculum Standards in an effort to improve future curriculum.

Table 18.2 Survey on students and teachers' attitudes across three subjects (percent)

	Studying this subject is important.		Studying this subject will be useful in my everyday life and my life in society.		If I study this subject, I will be able to think of ways to make society better as one of its members.	
	agree	disagree	agree	disagree	agree	disagree
World History	53.0	40.1	35.1	56.1	40.9	48.2
Japanese History	60.3	33.7	38.4	53.4	44.8	44.1
Geography	68.8	26.0	65.1	29.0	50.8	29.4

Source: National Institute for Educational Policy Research (2007)

Geography and History teachers were asked whether they were conducting "lessons that incorporated problem-solving learning," "lessons that incorporated developmental issues," and "lessons that made use of computers." The highest number of affirmative answers for all questions was for Geography. The results indicate that Geography teachers apply their expertise and use the largest variety of instructional methods to transmit knowledge.

Students' value of geography is also revealed in the survey. Senior high school students were asked to rank their attitudes towards studying by selecting one of the following responses: "agree", "somewhat agree", "somewhat disagree", "disagree", and "not sure" (Table 18.2). The results indicate that students who study Geography have a sense of its usefulness and reflect teachers' demonstration of geography expertise.

Although students do not rate the study of History very highly in terms of usefulness in their everyday life and their life in society, they do feel that history is important to help them improve society. In contrast, students believe that studying Geography is more useful in their own daily life and in society than it is in helping them improve society. Although Geography is recognized as offering knowledge and skills that are useful in daily life, it is difficult to persuade students to translate these into fostering citizenship, which is one objective of Social Studies education.

This survey affirms that geography education supports the development of responsible citizenship in students. Once students graduate from junior high school, they should recognize this as a goal of Geography. However, not all students understand the importance of citizenship. Therefore, finding a solution to this from the objective of Social Studies education is one of the greatest challenges facing geography education.

18.2.3 The Relationship Between Geography Education and History Education

For a long time, senior high school teachers taught geography under the Social Studies umbrella, without focusing on any one subject in particular. However, with the 1978 revision of the National Curriculum Standards, contemporary Social

Studies was mandated as a compulsory subject, changing the way Social Studies-related subjects were taught. Although not as large a reduction was observed with Ethics, there was a drastic decline in the number of students taking Geography, which had enjoyed the highest student enrollment until the change. Moreover, this low student number became definite with the 1989 revision of the National Curriculum Standards, which divided Social Studies into the two elective subject areas entitled Civics, and Geography and History, while making World History a compulsory subject.

While World History became compulsory, the vast majority of university applicants selected Japanese History, giving it a secure place within Social Studies. In 2006, there was a public outcry when it was announced that students were not taking World History course. The way that it was covered by the media pitted World History against Japanese History, but it never really addressed the dramatic changes that had taken place in Social Studies reform. Thus, the question of whether it was acceptable for students to not study World History until senior high school was never discussed.

World History became a compulsory subject because decision-makers believed it could play a role in fostering internationally minded qualities in the face of globalization. However, Geography was left out as a compulsory subject with potential to contribute to this student development.

When World History became compulsory in senior high school, teachers reacted by focusing on the history of Japan in junior high school. In essence, the history curriculum of Social Studies at this level informally became that of Japanese history. With this change, the topics taught in Geography shifted to focus on awareness about Japan's territory, and emphasis was placed on this knowledge at the junior high school level. In senior high school, the focus of geography shifted to global awareness. As a result of the compulsory subject change, junior high school students no longer learn about World History and they therefore receive a fragmented knowledge and understanding of world geography. Students will have encountered formal instruction of World History for the first time in senior high school and will not be able to depict a geographical image of the world.

Due to numerous factors, including global financial issues and unstable political situations, fewer young people desire to study abroad or be transferred abroad by their company. It is unofficially recognized that the young people of Japan today are inward-looking. The fostering of international-minded qualities may also have been negatively affected by the division of senior high school Social Studies into subcategories, as well as by the curriculum changes in junior high school Social Studies since World History was made a compulsory subject.

Revisions to the Fundamental Law of Education and the School Education Act strengthened the fostering of students' identity as Japanese citizens who love their country and hometown. However, such identity is relative. For example, within Japan, one may identify their roots by prefecture and city name. When the same individual travels abroad, their identity takes on a different scale and they are more likely to say that their roots are from Japan rather than to go into locational details.

World History remained a compulsory subject for senior high school students after the 2009 revision of the National Curriculum Standards. However, the importance of the relationship between different subjects was a fundamental idea that came out of the revision. Crosscutting concepts and topics that can be reinforced by a geographic and historical perspective should be supported, because both subjects benefit from an active effort to build a cooperative and complementary relationship.

Since the emergence of Social Studies as a discipline in 1947, it has undergone organizational changes at the senior high school level due to revisions in the National Curriculum Standards. The 1960 revision made each of the following subjects compulsory in senior high school Social Studies: Ethics and Contemporary Society, Politics and Economy, and Japanese History. For Geography and History, two subjects (Geography A and B) were set up, and students were required to select either A or B. In effect, every subject in Social Studies was compulsory. Leaving aside this period, the requirement to select one or two subjects from among those related to Geography and History was based on regulations governing compulsory subjects before 1989, the year that World History became compulsory. Over time, Geography- and History-related subjects have been offered in high school as elective subjects but are no longer compulsory. The decision to make World History a compulsory subject area was an unusual measure, even from a historical point of view. For example, Contemporary Society was first introduced as a compulsory subject although it has become an elective course under the Civics umbrella. With History, too, it makes sense to adopt an optional system for all subjects, from a historical viewpoint and in view of the deregulation of the National Curriculum Standards that is presently underway.

18.3 The Role of Geography Education

18.3.1 Geographical Thinking and Sustainable Society

What is geographical thinking? First, it is spatial thinking, namely, learning to position phenomena spatially and to understand them by freely using global, regional, and local perspectives. Second, it is to think in line with the facts of a specific region. Third, it is to understand the relationship between nature and humans. If we understand geography as a discipline that teaches students to consider how to make society better using such geographical ways of thinking, exactly what role should geography shoulder in today's society?

It is essential to learn from the past and to make judgments and undertake actions based on this. However, in the twenty-first century, we are faced with an entirely different set of circumstances compared with earlier times. If we follow the traditional road to growth, in which economic development is synonymous with social development, there is a pervasive lack of concern for matters such as maintaining a sustainable environment and society. We are facing a grave crisis. Pressing questions, such as how to sustain the earth's limited resources in our capitalist society and how to build a renewable-based society, form the essential themes of a future-oriented Social Studies education. To this end, Geography, as a part of Social Studies education, must strongly emphasize the relationship between nature and human society.

More than 10 years have passed since the dawn of the twenty-first century, which was dubbed "the century of the environment". The current international education trend is focused on fostering citizenship that includes global vision, with the aim of creating a sustainable society. Since 2005, the United Nations has developed a range of educational activities as part of the Education for Sustainable Development (ESD) initiative.

The National Curriculum Standards Revision report released in 2008 contained the phrase "aiming for the realization of a sustainable society" (Central Council for Education 2008, p. 79). In the 2008 National Curriculum Standards, Social Studies education in junior high school included content that encouraged students to consider actions and philosophies from "the perspective of forming a sustainable society" (MEXT 2008). This content is part of Civics standards, namely, aiming for a better society, and is positioned as a summary of Social Studies that includes the areas of Geography and History. This same perspective is featured throughout junior high school Social Studies. In fact, in Geography, Environmental issues and environmental conservation raises these issues as core ideas that are included together with content about the regions of Japan. In this topic, students are required to consider the importance of efforts towards environmental conservation for the purpose of building a sustainable society. Environmental issues and environmental conservation are core issues and [these efforts] should be tied back to the attitudes of industry, regional development, and people's livelihoods.

Senior high school Geography A includes a study unit on global issues in which students learn about the environment, resources, energy, population, food, housing, and urban issues from global and regional perspectives. This course makes clear the necessity for national initiatives and international cooperation to build a sustainable society that finds solutions to such issues (MEXT 2009). These objectives will form the central theme of future geography education.

Although the ultimate goal of Social Studies education is to foster citizenship through social awareness, Geography will remain a knowledge-based subject that contributes to student learning through phenomena and facts, which in turn leads to social awareness that directly contributes to the goal of Social Studies. Thus, Geography plays a critical role in developing scientific social awareness that forms the premise of value judgments and decision making for thinking about how to improve society.

The formation of a sustainable society requires a balance among environmental, economic, and social dimensions. Geography, which contains the elements of nature and humans, can develop convincing explanations about environmental dimensions based on the results of scientific inquiry. There is little doubt that geography can be a driving force of Social Studies because it includes ESD viewpoints such as the relationship between human activities and natural elements (e.g., topographic features, hydrologic features, climate, and terrain), developed by investigating case studies.

18.3.2 Reevaluation of Regional Geography Education in Social Studies Education: Focus on World Geography

As mentioned previously, geography plays an important part in developing scientific social awareness that forms the foundation of value judgments and decision making needed to improve society. For this purpose, regional geography, which studies specific phenomena and facts from an ESD perspective, has much to contribute. The following presents an ideal state of world geography.

The 2008 revision of the National Curriculum Standards introduced a regional study of Japan and the world to the geography portion of junior high school Social Studies. Regional geography education should not merely explore separately the characteristics, natural environment, industries, and culture of a region. Instead, the author believes it is essential to explore and construct, from an ESD perspective, specific case studies that reveal the lifestyles and ideas of people living in specific regions. In particular, a role of geography within Social Studies education is to provide an enriched world-level knowledge base. International content is well entrenched in geography instruction; however, it is less representative in the area of history, since Japanese History is the subject taken by most students. Given this, world-level regional geography education becomes essential to raising global awareness.

It is difficult to radically change the views of senior high school students, which have become somewhat entrenched by the time they reach high school. Thus, demonstrating the value of geography education in junior high school is especially critical to provide students with sufficient knowledge and perspective to form their own view of the world, such as understanding that the world has diverse cultures and environments that differ from those of Japan.

What is the best time to introduce international learning content? From the lower grades to middle grades of elementary school, there is a sense of accomplishment gained by simply remembering facts. Therefore, at this level, it would be effective to teach the names and locations of countries around the world, for instance, by linking them with national flags. What is essential in this practice is for students to: (1) regularly use maps and globes; (2) become familiar with maps from the lower grades; (3) focus on the distributions of continents, oceans, and the locations of countries; and (4) develop a spatial perspective.

For some time, the National Curriculum Standards has designated "The ways of life of people in Japan and people in countries that are deeply connected economically and culturally to Japan" as the final study topic for the sixth grade. It has been argued that world geography should also be taught from the middle grades onward. According to Iwamoto et al. (2010), this argument is based on the grounds that budding interest in and concern about foreign countries is seen in middle grade-level students. Children's sense of belonging (i.e., "I am Japanese") begins to develop from this period; once they reach higher grades in school, their ideas about specific foreign countries typically become fixed.

For a long time, the predominant view in geography education was that since children's spatial interests begin at a small scale (i.e., their local community), it is best for instruction to align with their cognitive development and start from the local community, moving to prefectures, to a discussion about the country of Japan, and then on to the world. However, there is no clear evidence that children's interests expand from local to international scales. Children's interests may actually be more indicative of having isolated knowledge of a place and that they have not learned about distant countries.

The literature does suggest that children are more interested in the outside world if they have some personal connection to it. Assuming that the scope of children's interests and enthusiasm expand beyond their immediate space, it should prove useful to introduce the study of a different community. In other words, in tandem with the study of the children's own local community, the curriculum can guide the exploration of a different community of a similar scale and to focus specifically on the community's phenomena that are familiar to the children, such as social activities and daily necessities (e.g., food, clothing, shelter and so forth). It is also desirable to provoke children's interest in the natural environment, which is the basis for people's livelihoods. This should be accomplished through studying regions where the environment is completely different from their own, without actually bringing up ideas of nations, but only by referring to regions as, for example, "southerly" or "cold". Students can follow up by considering the similarities and differences between the communities. Such teaching should also help children realize that they share things in common even with the people of faraway regions who are considerably different. If such an approach is taken, the best time to introduce regional geography is midway through elementary school, which sets the stage for children to learn about the world from this period onward.

The academic subject of Social Studies should not be held exclusively responsible for teaching international understanding. Instead, each subject under this umbrella should be capable of making a contribution through a cross-curricular approach. In this context, Social Studies would benefit from taking a regional geography approach. In other words, Social Studies can provide the groundwork required to study about people, culture, and place as an integral regional unit within a geographic context.

18.3.3 Building a Consistent School Curriculum with Regional Geography as the Core

In Japan, geography education is conducted as part of Social Studies in the broad sense. As this is unlikely to change in the near future, geography will continue to provide the foundation of Social Studies and play an integral role in the subject. As the basis for Social Studies, regional geography should be taught in elementary schools, which would facilitate an understanding of the lives of people in various regions and impact knowledge about the relationship between nature and society.

Specifically, in the middle grades of elementary school, comparative study should be made that examines the familiar phenomena (e.g., daily necessities, school life, and others) of a region similar in scale to children's local community. Through the comparative process, students will learn about the world beyond Japan.

I suggest that two case studies of such regional comparison be included in the curriculum. In the higher grades, it is important to continue a regional geography curriculum that focuses on the relationship between nature and human activities by studying about people, culture, and nature. Three case studies selected from among regions of Asia, Africa, and Oceania should be examined.

At the junior high school level, geography should reinforce world geography. I suggest a combined Geography and History type of regional geography approach that is rich in content about the natural environments, cultures, and historical backgrounds of each region. This should form the nucleus for global awareness instruction. In senior high schools, there should be instruction of a "combined Geography and Civics type" approach that is weighted towards Civics content (e.g., Politics and Economy), together with systematic geography that builds on what students have learned thus far.

Rather than repeatedly studying geographical descriptions, it is essential to develop an integrated approach that emphasizes different regions while introducing new perspectives as they arise.

18.4 Conclusion

If geography education pays little regard to natural elements and topography in favor of emphasizing the economic aspects of perceived phenomena, the subject will likely continue to be more focused on civics. However, such a trend would not be in line with the contemporary context of geography as part of Social Studies education.

As part of Social Studies at the elementary and junior high school levels, geography's prime objective should surely be to shift the outlook from economic to environmental aspects and to foster citizenship based on scientific awareness of the relationship between nature and human society. Based on the outcomes of physical geography and an accurate scientific understanding of the crises faced by today's global society, it is desirable to propose a clear vision of a "sustainable society" and to discuss what must be done to achieve it. In this respect, ways are needed to better understand how human, cultural, and regional aspects fit into Social Studies. This is accomplished by understanding the regional diversity in the world through the step-by-step teaching of world geography from elementary school onward, as well as by cultivating comparative thinking that combines the environment and society.

If geography education is to provide the groundwork for sustainable, future-oriented thinking and value judgments based on geographical knowledge and skills, teachers can be sure of the significance of geography and conduct their lessons with confidence. As for students, they will be able to engage in their study of geography with enthusiasm because the subject will give them the tools with which to build their sustainable "better society" of the future.

References

Arai M (2010) Seito ni totte manabigai no aru chishi gakushu towa: ikikata ya shakai no arikata wo kangaeru gakushu (What is regional geography education that is valuable for students? Education that discusses ways of living and how societies should be). Nihon syakaika kyoiku gakkai zenkoku taikai happyo ronbun-shu (Proc Jpn Assoc Social Stud Ann Conf) 6:258–261

Benesse Educational Research & Development Institute (2010) Dai 5 kai gakushu shido kihonchosa: Shogakko,Chugakko ban (Fifth basic survey of academic performance in elementary school and junior high school). Benesse Educational Research & Development Institute, Tokyo. http://berd.benesse.jp/berd/center/open/report/shidou_kihon5/sc_hon/index.html. Accessed 2 Dec 2013

Iwamoto H et al (2010) Shakaika chiriteki bunya ni okeru tangen "Mijika na Chiiki" no jissen-jokyo: zenkoku no chugakk shakaika kyoin taisho no anketo chosa wo tooshite (On "community area" geographical education in social studies at japanese junior high school: through the questionnaire research to social studies teachers). Nihon chiri gakkai happyo youshi-shu (Proc General Meet Assoc Jpn Geogr) 78:172

National Institute for Educational Policy Research (2007) Heisei 17 nendo kyoiku katei jisshi jokyo chosa (kotogakko) chosa kekka gaiyo: Chiri-rekishi (Results of state of the implementation of the national curriculum for senior high schools in 2005: geography and history). National Institute for Educational Policy Research, Tokyo. http://www.nier.go.jp/kaihatsu/katei_h17_h/index.htm. Accessed 2 Dec 2013

Chapter 19
The Current State and Prospects of Senior High School Geography Curriculum

Katsuki Toida

Abstract This chapter discusses geography education primarily at the senior high school level, which continues to be an elective subject even in the current version of the National Curriculum Standards. Since the 1989 version, geography has been relegated to the status of an elective subject which contributed to the enrollment decline. It seems that this trend will continue and it is inevitable that geography education in Japan will weaken further. A review of the issues confronting the Geography and History curriculum is presented. Further, the author proposes a history subject that is an amalgam of World and Japanese History, different from the current idea of reorganizing subjects. The envisioned four-subject curriculum structure includes the following: History of Humankind (four credits) and Modern History (two credits) which is a reorganization of traditional Japanese History and World history, General geography (four credits) and Basic geography (two credits), which are based on traditional Geography A and B.

Keywords Curriculum • Geography and History • Senior high school geography • The National Curriculum Standards

19.1 Introduction

The new National Curriculum Standards was announced in March 2008 for elementary and junior high schools, and in March 2009 for senior high school. These National Curriculum Standards have been implemented since April 2011 in elementary schools, April 2012 in junior high schools, and April 2013 in senior high schools. The goal of the chapter is to discuss issues surrounding the geography curriculum in senior high school and to propose prospective future directions.

K. Toida (✉)
Kinki University, 3-4-1 Kowakae, Higashiosaka, Osaka 577-8502, Japan
e-mail: toida@socio.kindai.ac.jp

Y. Ida et al. (eds.), *Geography Education in Japan*, International Perspectives in Geography 3, DOI 10.1007/978-4-431-54953-6_19

A report of the Central Council for Education (January 2008), which included criteria for the new National Curriculum Standards, suggested the following seven points for its improvement (Central Council for Education 2008):

(1) Revising the National Curriculum Standards based on the revised Fundamentals of Education Law;
(2) Sharing the idea of "abilities in living";
(3) Acquiring basic and fundamental knowledge and skills;
(4) Developing abilities to think, judge, and express;
(5) Securing the number of class hours that is required to establish academic abilities;
(6) Increasing learning motivation and establishing study habits; and
(7) Enhancing instructions to nurture spiritual richness and physical health.

These criteria address fundamental principles in general education as applied to elementary, junior and senior high schools. In considering the content of geography curriculum, items (3) and (4) are especially important. Item (3) has implemented a shift from traditional case studies to systematic topographical studies in geography instruction. This is in addition to the traditional instruction to stress geographical perspectives, ways of thinking, and nurture application abilities. Item (4) is implemented by accentuating students' linguistic abilities, which includes the core abilities to think, judge, express opinions (e.g., written and oral), and facilitate voluntary "social participation". Social participation is emphasized by the new provisions of the revised Fundamentals of Education Law (December 2006), which is referred to in Item (1), and the revised School Education Law (June 2007). The idea can be interpreted as recognizing the essential roles of Social Studies. In summary, the revisions steered significantly away from "knowing how" to emphasizing "knowing what."

19.2 State of Senior High School Geography Curriculum

Today, geography education in senior high school is in critical condition. The initiation of the Geography and History subject in 1989 accompanied by the reorganization of Social Studies. As a result, World History became the only subject under the Geography and History umbrella to become a compulsory subject. The other three subjects, Geography, Japanese History, and World History became elective courses. This policy change has partly triggered a significant drop in Geography enrollment. The elective status of geography remains unchanged in the curriculum revision of 1999 and in the new curriculum of 2009. Geography enrollment has steadily declined, and in the author's experience, for example, only 10–20 % of college students today have studied Geography in senior high school. For nearly three decades, equivalent to an entire generation of young people, they will very likely reach adulthood without the necessary knowledge and understanding of geography.

This is a serious problem not only for learners but also for teachers. In junior high schools, Geography is required, and there is a need for many geography instructors, however, Social Studies instructors report that Geography is not their strength. According to a survey conducted by the author, who has been involved in teacher training courses, the number of subjects that students take for college entrance exam within specialty areas in Social Studies was higher than that taken by majors in college. In "Geography and History, Civics, and Math," which is one of the entrance exam subjects at the author's university, Japanese History is the most selected exam, accounting for nearly half the students. The remaining half of exams is shared by World History and Politics and Economy, with a small proportion shared by Geography and Math. In fact, the majority of the students who are enrolled in the teacher training course taught by the author wish to teach Japanese History despite the fact that they are enrolled in departments of law, economics, and management. However, it is foreseeable that these students will be assigned to work in junior high schools and teach Geography, without any university training in this subject area.

One of the chief factors that contribute to enrollment decline in senior high school geography is that geography is not an option for many college entrance exams, especially private colleges.[1] A document edited by the AJG-CGE (Association of Japanese Geographers-Commission of Geographical Education)[2] reveals that out of 860 departments/programs and 273 private universities, 848 (98.6 %) of departments and programs offer World history as an option, and 810 (94.2 %) offer Japanese history, while only 380 (44.2 %) offer Geography. Private institutions are public education organizations that receive subsidies, so they are they not socially responsible to accept entrance exams from subjects' organization in the National Curriculum. Since Geography and History consists of three areas, World History, Japanese History, and Geography, from a macroscopic point of view, an acceptance structure of entrance exams that unjustly neglects one of these areas can be considered to be against the national educational policy.

World History was made into a compulsory subject based on rationale not extended to other Geography and History subjects. This change along with the insufficient number of geography teachers in junior high schools, and the selective acceptance of Geography for college entrance exams are all interrelated and have weakened senior high school Geography. This negative spiral that is happening may ruin the geographical culture within the Japanese society.

---

[1] Out of 773 4-year colleges in Japan, 86 are national universities (about 620,000 students), 92 are public universities (about 140,000 students), and 595 are private universities (about 2,090,000 students). The number of private universities and their students account for an overwhelmingly high percentage of the total (MEXT 2010).

[2] Report by the committee members, Shigeru Sugiki and Masami Komiya as of December 2010. Incidentally, because combinations of possible subjects differ depending on colleges, detailed statistics need further investigation.

19.3 Overview of the New Geography Curriculum

The following section provides an overview of the new senior high school geography curriculum. In recent years, the National Curriculum Standards has been composed of "objectives," "contents of learning," and "how to deal with the contents of learning." Objectives represent the fundamental idea of geography, the contents are in-tended to embody the idea in the objectives, while the section on how to deal with the contents is what teachers need to consider when instructing the topics.

19.3.1 Objectives

New objectives were progressively developed by revising previously published materials, which reflect social change and scholarly trends since the last National Curriculum Standards. A list of the objectives from the new Geography curriculum is presented below, with the new revisions underlined.

- [Geography A] Objectives: Consider geographical issues in the modern world in relationship with regional perspective, historical background, and daily living; develop geographical knowledge on the modern world; cultivate geographical perspectives and ways of thinking; and foster awareness and characteristics as Japanese citizens who live in the global society in independent-minded manners.
- [Geography B] Objectives: Consider geographical events of the modern world in a systematic geographical way and regions in the modern world topographically accounting for historical background; develop geographical knowledge on the modern world; cultivate geographical perspectives and ways of thinking; and foster awareness and qualities as Japanese citizens who live in the global society in independent-minded manners (MEXT 2009).

While following closely to the previous version, the current revisions are distinctively characterized by their emphasis on historical background as a dimension of thinking about the world. The term 'historical background' was formerly included in the contents section, but it has been included in the objectives for the first time emphasize its significance. Questions have been raised about the lack of historical considerations in geography, which dealt primarily with the modern world, and whether students should have a more dynamic perspective to understand evolving geographic issues across regions. Key terms changed in the course of revision which are elaborated on below.

The term "Japanese citizens" in the revised curriculum replaced the word "Japanese" in the former version. This change is parallel to the revision made in the Fundamentals of Education Law (December 2006), a turning point for the National Curriculum Standards. The term "daily living" is used only in Geography A to reflect the nature of an "A" subject. Common to World History, Japanese History, and Geography, "B" subjects are assigned four credits and comprehensively covers contents in the traditional way. In contrast, A subjects are assigned two credits and are considered a condensed version of version B subjects. World History and Japanese History A courses have accomplished this by stressing on modern times, while Geography emphasizes on geographical issues to accentuate the differences with Geography B. The emphasis on daily living is in line with these trends. In the contents section, the concept of "disaster prevention" was introduced in this version for the first time, stressing more on area surveys than before. As a result, the new learning content, "disaster prevention", was mentioned clearly in Geography A,[3] instead of Geography B.

19.3.2 Contents and Treatments of Contents

The following are characteristics of the new Geography curriculum content:

(1) Learning that is connected with the real life (especially Geography A);
(2) Enhanced learning of world topography (especially Geography B);
(3) Introduction of inquiry learning;
(4) Emphasis on abilities in resource application and expression; and
(5) Application of atlas.

Items (2) and (5) are examined further here. Item (2) signifies in a practical sense the abolition of case studies, which was a measure adopted when the National Curriculum Standards was announced in 1989. Formerly, instruction on topography was not valued, and instead, teachers focused intensely on two to three countries or regions, using them as representative case studies. The rationale was that intensive learning about particular countries and regions would help students develop the academic abilities to independently generalize about other countries and regions. This was part of the curriculum that focused more on "knowing-how" rather than "knowing-what". In response to students' declining basic academic abilities, the policy was changed to emphasize "knowing-what". In principle, the acquisition of geographical knowledge, perspectives, and ways of thinking should be accomplished by inductive information processing, that is to find general commonalities through multiple examples. Therefore, it is natural that the acquisition of geographical knowledge tends to be a comprehensive study. On the other hand, the deductive learning method, based on two to three representative countries or regions, is not consistent with current models of geography learning.

Item (5) reaffirms the role and status of the atlas as a textbook. The atlas used in senior high schools is titled "Atlas for senior high school Geography and History: Authorized textbook by MEXT". It has the status of a textbook and is used in Geography and History. Fundamental changes have made this an atlas for Geography and History, both in name and reality, rather than solely in Geography instruction.

[3] The National Curriculum Standards commentary suggests that "disaster and measures" may be addressed as a part of survey content in regional survey in Geography B.

The use of the atlas in Geography and History is specified in the National Curriculum Standards, the result of efforts by many stakeholders in geography education and the AJG-CGE who are concerned about the critical state of geography. Geography instructors should be more appealing as the atlas is a useful textbook relevant to the entire subject of Geography and History, as well as World History, and Japanese History.

19.4 Necessity for Senior High School Geography

The immense undervalue of Geography is unprecedented in the history of geography education in modern Japan and in other foreign countries (e.g., European countries, Korea, and China) in the recent years. The state of geography education is critical and a lack of action is a sign that the Japanese society agrees that senior high school geography is not needed. Is geography really obsolete in senior high school? Needless to say, the answer is no. To the contrary, geography is going to be even more critical in the future, and this is discussed from two perspectives in the next section.

19.4.1 The Importance of Geography Education from the Perspective of the Curriculum of Geography and History

The curriculum of the senior high school Geography includes history concepts while history education also encompasses geography. This is evident in the new History curriculum. Objectives of World History A and Japanese History A from the new National Curriculum Standards are quoted in the following, with new revisions underlined:

- [World History A] Objectives: Based on resources, help students to understand world history focusing on modern times in relationship with <u>geographical conditions</u> and Japanese history; cultivate abilities to think historically <u>by helping students to consider modern issues from historical perspectives</u>; foster awareness and qualities as Japanese citizens who live in the global society in independent-minded manners.
- [Japanese History A] Objectives: Based on resources, help students to understand the history of modern Japan in relationship with <u>geographical conditions</u> and world history; cultivate abilities to think historically <u>by helping students to focus and consider modern issues</u>; foster awareness and qualities as Japanese citizens who live in the global society in independent-minded manners.

First, the term, geographical conditions, was used in the contents section, as was the term, historical background, in geography. Moving it to the objectives section represents a greater emphasis on it. The underlying emphasis of geographical conditions in history courses suggests a focus on the regional natural environment or spatial relationships in understanding history with the intent to teach viewpoints and methods of history with an atlas.

The notable term modern issues underlines the effort in history curriculum to discuss important current topics as they arise. According to the National Curriculum Standards, the ultimate objective of World History and Japanese History is to foster Japanese citizens who live independently in the global society. This goal is common in both A and B subjects of World History, Japanese History, and Geography. Cultivation of historical thinking abilities is a necessity and the integration of modern issues is one method to support this development. The inclusion of modern issues into history is implemented into A and B subjects. A modern issue is a loss of self-sufficiency in food production, owing to desolate agriculture, changing fishing villages, and a loss of traditional culture. In World History, these current world issues are couched in the discussion of north–south problems that include intense ethnic conflict and a loss of biodiversity. Both examples have been studied or are likely to be studied in Geography. By itself, how would history education address these issues? It is time that Geography and History work in unity to the degree necessary to grow. A geographical perspective and ways of thinking (e.g., concepts of region and space, natural environment unique to the region), will provide useful perspectives in the investigation of modern issues as well as in other elements of history, such as examining causality and rationality in ancient and medieval history.[4]

19.4.2 From the Perspective Geography Education Goals

Geography has unique perspectives and characteristics that can easily be transferredable to related subjects. Three examples are provided. First, as a science of space, geography is concerned with the scale of the region being discussed. In particular, region and space have various scales, such as the local scale where people live, to the national scale, and the space outside Earth. These topics are taught in context of location and space in geography whereas the course Contemporary Society in Civics provides only a cursory discussion of scale. Thus, the slogan "Think globally, act locally," is a fitting argument for geography, where scale matters.

Second, Geography is able to provide multiple perspectives on phenomenon at interlocking scales. For example, the self-sufficiency issue on food is often discussed from the standpoint of food security and securing safety in food. The views from Politics and Economy are important in consider. However, from a geographical

[4]The senior high school National Curriculum Standards of 1960 and 1970 mandated that four subjects or more including Ethics and Civics as well as Politics and Economy were required out of the following five: Ethics and Civics, Politics and Economy, Japanese History, World History, and Geography. In reality, all five subjects were made compulsory in the majority of general senior high schools.

perspective, the essence of the problem emerges from multiple and diverse factors that are different at various scales. At the hometown scale, a breakdown of agriculture and fishing villages and the accumulation of organic waste are examples while human-made reservoir dams are at the national level. This is compounded by issues at the international level where developing countries lack steady food supplies and a spike is in carbon dioxide due to increased food mileage from transportation. These issues are interrelated across hometown, nation, and worldwide scales, making the essence of the problem even more complicated. Geography can provide multi-faceted and diverse ways of thinking about these complex issues.

Third, it is important to examine issues from the standpoint of "interaction between nature and human" at all scales. Considering the recent circumstances of the world, the questions such as how humans should face nature and what are sustainable developments for society are increasingly important. Nonetheless, the standpoint of "interaction between nature and human" is still decidedly absent in not only Civics but also in Japanese History, and World History courses. This perspective is not limited to development and environmental preservation. In the world, there are ethnic groups whose social styles, values, and life cultures are diverse, a reflection ofthe natural environment that is their home. Even in the twenty-first century, there is discord among ethnic groups, and quarrels are not resolved. Harmony and integration should be approached from temporal, historical, and conceptual standpoints as well as from a natural, special-compositional, and existential views.

These views and ways of thinking are the same as those in geography as presented in the National Curriculum Standards. The integration of humanities and science is going to be in demand in the world of the twenty-first century, and geography education can respond appropriately.

19.5 Curriculum Strategies for Regeneration of Geography Education

It is imperative to return senior high school Geography to the state prior to 1989 when there was balance between Geography and History education. Policy-related prospects for that vision are presented below.

Presently, basic research on how to best integrate Geography and History is commissioned by MEXT and conducted in some research development schools. The research has only just begun but it is possible that the findings may be reflected in the next version of the National Curriculum Standards. The idea of reorganizing Geography into a new integrated study, such as environment, international understanding, and area studies, has been replaced by a new movement to merge geography and history subjects into an integrated course. Tentatively, the suggested course title for the merged course is Integrated Geography and History. Alternatively the current course may be restructured to Basic Geography and Basic History.

19.5.1 *Double-Feature Programming of Basic Geography and Basic History*

There has been an idea to establish a subject that integrates geography and history, to be called Integrated Geography and History (tentative title). This initiative is similar to the action chosen in early Social Studies. For example, the second National Curriculum Standards announced after World War II that in addition to General Society (five credits) as a compulsory course in the first years of senior high school, students would take at least one from the following four subjects: Japanese History, World History, Human geography, and Current Issues (five credits each). Moreover, by liberally incorporating historical components in Geography and geographical components in History, students still learned history, thus beginning a complementary relationship between Geography and History. In Human Geography, the following terms were purposefully integrated: exploration of Africa by Livingston and Stanley, the travel journal of Marco Polo, times when Mongolians have flourished, history of contact between Mongolian and Han people, industrial revolution, land reform, Malthus' Essay on the Principle of Population, castle towns, development of cities in medieval Europe, *Kojiki*, Columbus, Magellan, *Gishi Wajin-den*, Roman roads, and the Silk Road. This provides a frame of reference in the new curriculum that emphasizes historical understanding.

However, under the current curriculum, it is difficult to carry out a radical change as in the past because the number of existing geography instructors is incomparably smaller than that of 20 years ago. If compulsory Integrated Geography and History is established under these circumstances, it is inevitable that instructors who specialize in History will be more likely to teach the subject. The classes will likely be far from what Integrated Geography and History should be, that is to say, one that weaves the human world with warp (history) and weft (geography). If this plan is accepted as an Integrated Geography and History subject, problems with the content remain.

Thus, he National Curriculum Standards' strategies to regenerate senior high school Geography must feature Basic Geography and Basic History (tentative titles). Furthermore, they must be made compulsory, even if assigned fewer credits to put Geography and History on equal footing.

19.5.2 *From World History and Japanese History to History of Humankind and Modern History*

The Geography and History course is divided into World history, Japanese History, and Geography. The issue with this is that geography only makes up a third of this division, with history dominating more teaching time. This is confounded by the illogical division of history into distinct topics of world and Japanese history. The current division of the Geography and History course into one part geography and two parts history (World and Japanese) is an imbalance that favors history. Another example

Table 19.1 High school Geography and History curriculum (Tentative plan)

Geography	History
Basic geography (two credits)	Modern history (two credits)
General geography (four credits)	History of humankind (four credits)
Requiring more than two credits	Requiring more than two credits

of a bias towards history courses is rooted in the teacher training system. For example, the Teaching License Law stipulates that students take the same number of geography and history credits regardless of their chosen teaching subject. Since the course Geography and History is divided into Geography, Japanese History and World History topics, it means that two third of instruction time are dedicated to history-related lessons. The author argues that if teacher candidates must study equal credits for geography and history, then more time for geography is needed in senior high schools.

The first National Curriculum Standards introduced after the war in 1947 divided world history into Eastern History and Western History. The second version in 1951, integrated these two topics into World History. Conversely, history departments at universities continue to make a distinction between eastern and western histories. This framework that separates the teaching of Japan's history from the rest of the world isolates events in Japan and it is thus taught in a vacuum, separate from the world context. In the U.K. as well, History is a concept that combines world history and British history. In addition History and Geography have been treated equally. In France too, history is a concept that combines history of the country and the world, and by positioning it alongside Geography, the axis of time and space are maintained.[5]

Here the author would like to propose a personal idea: the current World History and Japanese History be dismantled in favor of reorganizing History of Humanity (four credits) and Modern History (two credits). By moving between Japanese History and world History, both subjects will ideally highlight the past and challenges of humans. The author wonders whether the latter subject could be content-structured to trace the historical process in the discussion, while stressing challenges. However, could geography be incorporated into General geography (four credits) based on the current Geography B and Basic Geography (two credits) based on Geography A (all titles tentative)? Moreover, a strategy is needed to makes two credits or more in Geography and two credits or more in History compulsory from the above four subjects. Table 19.1 presents an early idea of the proposed course reorganization.

In the current Geography and History curriculum, which was launched in 1989, it has been divided into World History, Japanese History, and Geography. As a result of World History becoming compulsory, while Geography became elective, geography enrollment drastically declined. Consequently, geographical knowledge among

[5] National Institute for Educational Policy Research (2000, 2004) reports systematic research on curricula of subjects in Social Studies in foreign countries.

the young generation is considerably lacking, and balanced knowledge and culture between geography and history have significantly deteriorated. Fundamental reform in senior high school history subjects is necessary for the advancement of geography education.

References

Central Council for Education (2008) Yochien, syogakko, chugakko, kotogakko oyobi tokubetsu shien gakko no gakushu shido yoryo tou no kaizen ni tsuite:toshin (Improvement of the national curriculum of kindergartens, elementary schools, junior and senior high schools, and special support schools : report). January 17, 2008. http://www.mext.go.jp/b_menu/shingi/chukyo/chukyo0/toushin/__icsFiles/afieldfile/2009/05/12/1216828_1.pdf. Accessed 2 Dec 2013

MEXT (2009) Koto gakko gakushu shido yoryo (National curriculum for senior high school). Higashiyama shobo, Kyoto

MEXT (2010) Monbu Kagaku Toukei Youran (Statistical abstract of education, culture, sports, science and technology). MEXT, Tokyo

National Institute for Educational Policy Research (2000) Syakaika-kei kyoka no curriculum nokaizen ni kansuru kenkyu: Syogaikoku no doko (Studies on improvement in curriculum of social studies: trend of foreign countries). National Institute for Educational Policy Research, Tokyo

National Institute for Educational Policy Research (2004) Syakaika-kei kyoka no curriculum no kaizen ni kansuru kenkyu: Syogaikoku no doko (2)(Studies on improvement in curriculum of social studies: trend of foreign countries (2)). National Institute for Educational Policy Research, Tokyo

Chapter 20
The Great East Japan Disaster and Geography Education: School Damage, Associations' Support Project, and Trends in Geography Teaching Practices

Takashi Shimura and Kotaro Yamagata

Abstract The Great East Japan Disaster that struck on 11 March 2011 had a massive and wide-ranging impact on all aspects of life. It directly affected schools in the area and indirectly impacted geography associations, education policy, and geography education. This chapter provides a brief explanation of the multifaceted disaster and its aftermath, including earthquakes, tsunamis and the nuclear power plant accident. The authors' visits to and investigations of affected schools found that needs to support geography teaching post-disaster were dependent on the disaster(s) that struck, school locations, and level of destruction. To respond to these needs, the Association of Japanese Geographers in conjunction with other associations conducted emergency fundraising. At the government level, the disaster prompted discussions on natural hazards prevention education, as demanded by policy makers at the national level. Geography education is considered one of the main subjects of the disaster prevention curriculum. As a result of public demand, disaster prevention practices have notably improved since the event. For example, in the immediate days after the disaster, many lessons were given about its occurrence and processes, along with preventative education to foster students' knowledge and skills for surviving future natural hazards. Recently, thanks to interest generated by these lessons, geography classes have been increasingly concerned with best methods to restore devastated areas and to create a Japanese society resistant to disasters.

Keywords Affected schools • Disaster prevention education • Education for Sustainable Development • Fundraising for geography teaching resources • Nuclear disaster

---

T. Shimura (✉)
Joetsu University of Education, 1 Yamayashiki-machi, Joetsu, Niigata 943-8512, Japan
e-mail: shimura@juen.ac.jp

K. Yamagata
Joetsu University of Education, Niigata, Japan

Y. Ida et al. (eds.), *Geography Education in Japan*, International Perspectives in Geography 3, DOI 10.1007/978-4-431-54953-6_20

20.1 Purpose

A massive earthquake occurred offshore of the East Japan Pacific Coast on 11 March 2011. It caused a tsunami that in turn inflicted damage on East Japan and the surrounding area. This natural hazard is referred to as the Great East Japan Disaster. The areas affected were struck by at least one of three phenomena, earthquake, tsunami and/or nuclear power plant accident. Current restoration efforts to rebuild infrastructure and to create a new society are pressing challenges in Japan, one to which geography education can contribute.

This chapter addresses the impacts of the Great East Japan Disaster, addressing severely damaged schools in Iwate, Miyagi, and Fukushima. This is followed by an examination of emergency support to the damaged schools by academic geography associations and related organizations. Finally, current and post-disaster Japanese education policy and research trends in geography education are reported.

20.2 Characteristics of the Great East Japan Disaster

20.2.1 Overview of the Disaster

On 11 March 2011 at 14:46 JST (05:46 UTC), the plate boundary fault along the Japan Trench slipped over a length of approximately 500 km off the coast of Miyako City, Iwate Prefecture, extending to Chousi City, Chiba Prefecture. This generated a 9.0 magnitude (MW) earthquake, which was the most powerful known earthquake ever to have struck Japan. The number of deaths is estimated at 15,879, with a further 2,700 people still missing (National Police Agency Publicity 2013). The cost from direct damage was estimated to be 17 trillion JPY, which is approximately 20 % of the national budget of Japan (Tanaka 2012).

The disaster was characterized by a large-scale earthquake, which generated a large tsunami that brought about the majority of the damage. The distribution of dead and missing people was concentrated along coastal municipalities, suggesting that most of the damage was due to the tsunami (Fig. 20.1). The earthquake impacted a wide area including Tohoku and Kanto regions and its dangerous impact multiplied when the earthquake and tsunami caused the nuclear power plant disaster.

In response to the earthquake disaster, the word "unexpected" is often used. For example, Taro district in Miyako City, Iwate, was known for its most advanced structural countermeasures against tsunami. Double 10 m high tsunami seawalls extending over 2.4 km length were laid out throughout the district. Despite the structural protection, the tsunami flowed over the double levees and devastated Taro district. The seawall was constructed to protect against tsunamis of this exact size that struck this area during the Meiji and Showa eras. However, the seawall did not stand in the face of this "unexpected" tsunami, which was larger than those historically experienced in the area.

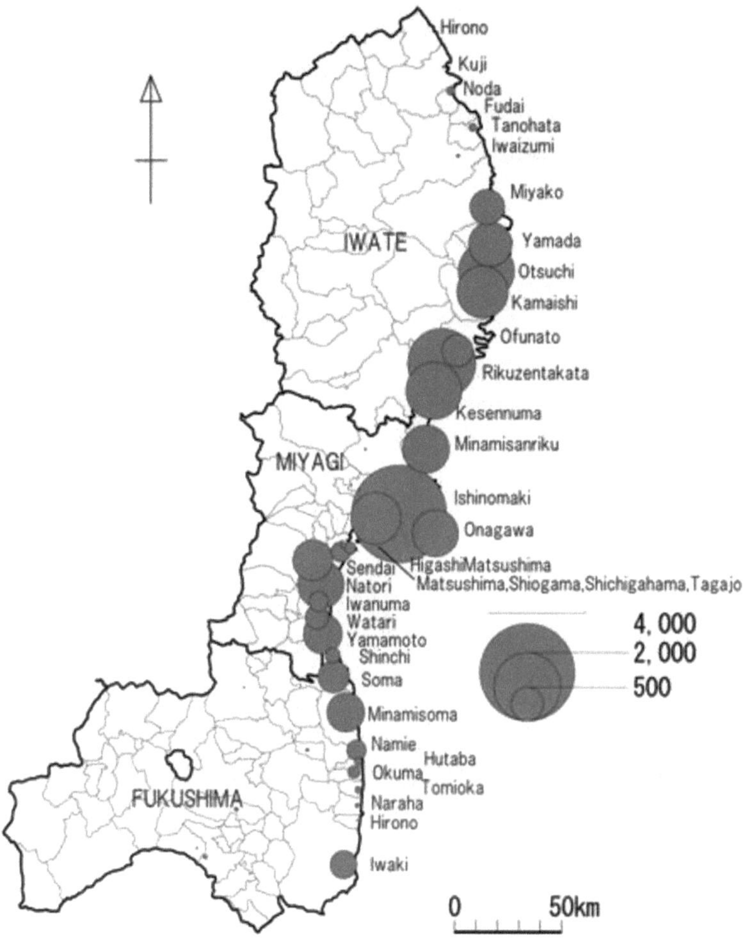

Fig. 20.1 Number of missing and dead within each municipality (National Police Agency Publicity 2011)

20.2.2 Earthquake and Tsunami Hazard

Plate boundary earthquakes of about M7–8 have repeatedly occurred off the coast of the Tohoku region (Obara 2012). Focal regions of these earthquakes are separated into eight zones. It was previously believed that each zone experienced such earthquakes at an interval of a dozen to several hundred years (the asperity model). However, in the case of the 2011 earthquake, six of the eight zones slipped together, a phenomenon that had not been expected in this region. The plate boundary displaced more than 50 m, resulting in the seafloor being vertically uplifted by 5 m. On land, it was also confirmed from GPS data that the Oshika Peninsula moved

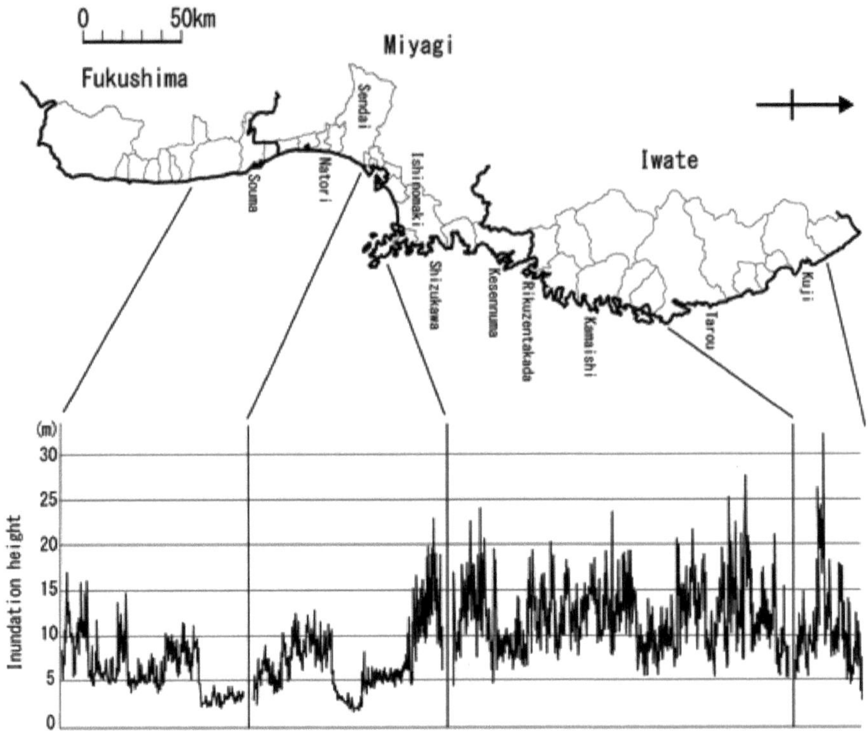

Fig. 20.2 Inundation height distribution of the tsunami (modified after Tsunami Damage Mapping Team, Association of Japanese Geographers 2011)

about 5 m in an east-southeast direction. This movement spread to eastern Japan, including all areas south of the Tohoku region.

The large vertical seafloor uplift generated a giant tsunami that was larger than expected by scientists. A tsunami damage mapping team of the Association of Japanese Geographers (AJG) developed a map of the devastated area from aerial photos, which showed the distribution of the inundation height (Fig. 20.2). The magnitudes of height fluctuations were large in the Sanriku coastal area, with a maximum inundated height of 30 m, attributable to the effect of a deeply indented coastline. In the regional plains between Sendai and Ishinomaki, inundation height and magnitude of fluctuations were small, owing to the effect of flat terrain. However, the inundation distance extended several kilometers inland, resulting in a wide area of flooding. Further south in Fukushima Prefecture, the inundation height was large because of the effect of coastal terraces.

The largest damage from the tsunami was not in the Sanriku coastal area but on the Ishinomaki and Sendai plains. A large population in this area meant that more people were at risk. Moreover, the tsunami inundation area greatly exceeded the assumed tsunami danger zone outlined in hazard maps created before the earthquake. The importance of these maps was highlighted after the disaster.

Susceptibility of the Ishinomaki and Sendai Plains was found in a recent geological survey that revealed a historical tsunami 1,000 years ago caused inundation as far as several km inland (Sawai et al. 2007). The Jokan earthquake tsunami of 869 AD caused similar damage to the inundation area of the 2011 tsunami. Unfortunately, these discoveries were not previously included in disaster-prevention measures before the March 2011 disaster.

20.2.3 Nuclear Power Plant Disaster

A nuclear power plant accident at Tokyo Electric Power's Fukushima Dai-ichi Nuclear Power Plant happened as a direct cause by the earthquake and tsunami. Large amounts of radioactive material were released into the environment, forcing residents who lived near the plants to evacuate, and many have not been able to return. Radioactive pollution in the environment spread across a wide area and industries in the affected areas have suffered from detrimental effects of fear and harmful rumors. For example, customers refuse to buy agricultural products from Fukushima Prefecture, even though crops are grown in low radioactivity areas or are produced in indoor vegetable factories. The name Fukushima raises alarm bells for buyers although only a small part of the prefecture is contaminated.

Investigations into the cause of the nuclear accident are ongoing. Current reports state that a tsunami of 14–15 m struck the nuclear plant, causing loss of external power supply. As a result, the emergency backup diesel generator was soaked in seawater and broke down, producing a blackout at the plant. It was impossible to cool the nuclear reactor, so it was assumed that nuclear fuel was melting. Hydrogen generated for energy became fuel for explosions, which dispersed a large quantity of radioactive material. This quantity was estimated at $1.5 \times 1{,}017$ Bq of I-131 and $1.2 \times 1{,}017$ Bq of Cs-137, dispersed between 11 March and 5 April 2011 (Great East Japan Earthquake Taskforce 2011).

20.3 School Damage in Three Affected Prefectures

20.3.1 School Damage Investigation

Many schools suffered damage in the earthquake disaster. As of 26 January 2013, the total number of deaths associated at schools was 641 and the total missing was 92, including students and teachers. Most were in Iwate, Miyagi, and Fukushima prefectures, with the largest number in Miyagi. A total of 6,284 school buildings were damaged (Kurokawa 2012).

In response to this grave disaster, the AJG decided to provide assistance to schools in the affected areas. This support included the provision of teaching materials and tools related to geography instruction at elementary, junior high, and senior high schools. To furnish relevant support, it is necessary to understand needs in the affected areas. The authors received a budget for emergent research programs on the Great East Japan Disaster, under the theme "The Clinical Approach for Developing Support Measures for Geography Classes During the Rehabilitation and Reconstruction Process after the Great East Japan Disaster[1]" given by the Tokyo Geographical Society (TGS). The research was led by the authors, who began by collecting information about affected schools with assistance from teachers in these areas. Field surveys were conducted whereby the authors visited the affected sites in May, June, and September 2011. Observations at the schools showed that each area learned from previous tsunami disaster and they had protective defences already in place. For example, regular practice drills at schools to construction of tall wall around villages are common. For example, many schools in the Sanriku coastal area are on hills. Given the physical geography, many students and teachers were able to escape the tsunami by going to higher elevation; however, schools in lowlands were the most affected.

20.3.2 Damage to Schools

Schools impacted by the tsunami were identified using the tsunami inundation area map developed by the tsunami damage mapping team. Figure 20.3 maps the tsunami-affected schools and the nuclear power plant accident in Fukushima Prefecture. The total number of affected schools was 101 in 34 municipalities (65 elementary, 25 junior high, and 11 senior high schools). Damage by geography comes to 30 schools in Iwate Prefecture (20 elementary, 7 junior high, and 3 senior high schools), 64 in Miyagi Prefecture (41 elementary, 16 junior high, and 7 senior high schools), and 7 in Fukushima Prefecture (4 elementary, 2 junior high, and 1 senior high school). Damages were especially severe in the plain areas, such as Ishinomaki and Sendai City (Fig. 20.3). Of the estimated 617 students who perished across Iwate, Miyagi, and Fukushima prefectures in the Tohoku region, Miyagi lost 430, or 70 % of the total student population. It was confirmed that many municipalities and schools were to be supported in these three prefectures. Schools in Iwate and Miyagi prefectures have reopened despite a loss of teaching materials. Restoration of schools is progressing, but those within the evacuation zone in Fukushima Prefecture remain closed. Satellite schools temporarily replaced the evacuated high schools. However, most elementary and junior high schools could not continue and function normally, because residents and students had voluntarily left the region. Almost 2 years after the earthquake disaster, the low student population has not changed.

Huge seawalls had been constructed at great financial expense in the Sanriku coastal area before the disaster. Those seawalls were supposed to protect schools in

[1]The program team includes K. Yamagata (head), T. Shimura, Y. Umetsu (Miyagi Prefecture Sendai Minami High School), T. Hanzawa (Iwate Prefectural Board of Education), H. Nagaike (Fukushima Prefectural Futaba High School).

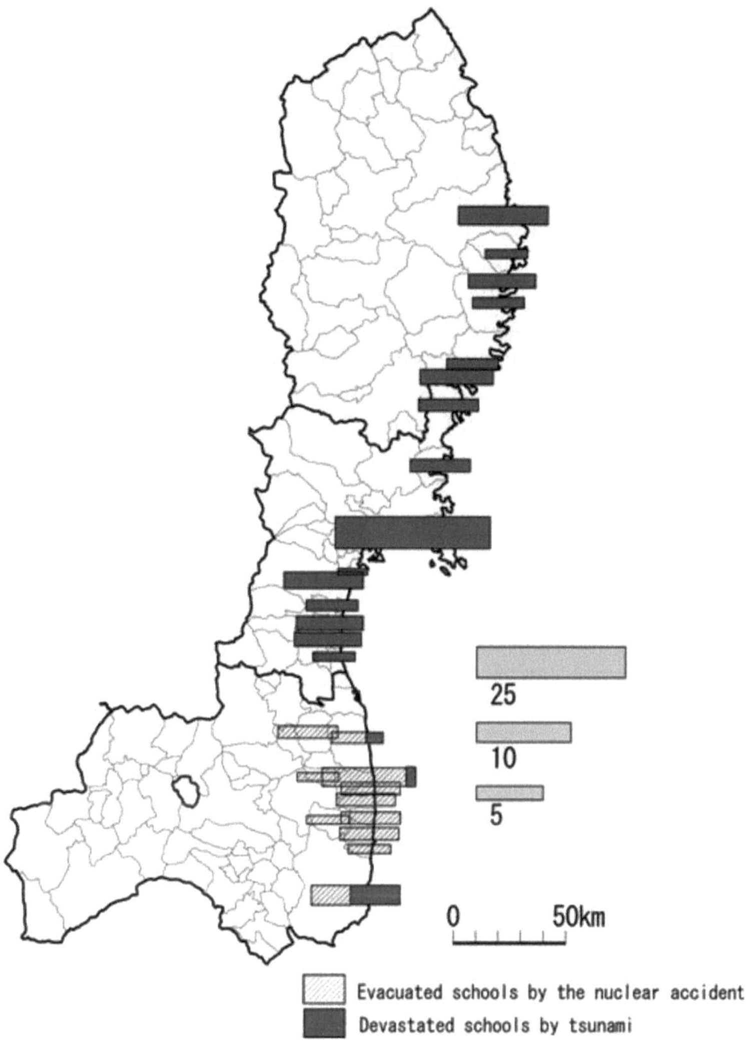

Fig. 20.3 Number of schools affected by tsunami and the nuclear power plant accident

lowland areas, but they collapsed. Thus, there is a limit to the protection that artificial structures can provide against natural disasters.

At the time of the earthquake, students at some junior and senior high schools began evacuating by moving to higher ground and narrowly escaped the disaster. Unfortunately, other schools suffered countless casualties because it took too long for the evacuation decision to be made. These cases demonstrate a real need for students to have the ability to save themselves and act autonomously, instead of waiting for instructions at the time of a disaster.

20.4 Emergency Responses by Geography Education-Related Associations and Organizations

Considerable support came from many people and organizations immediately after the earthquake. Japanese geographers belonging to the AJG, geography education-related associations and organizations and academic societies quickly organized various support activities.

On 14 March 2011, the Disaster Response Headquarters was led by the chairperson of the AJG. This headquarters coordinated rapid dissemination of information, including publicity with geographical disaster information to other academic societies and mass media. For example, instead of canceling the general AJG conference planned for 29–31 March in Tokyo, the head office held an emergency opening conference "Restoration from the Great East Japan Disaster and Geography" on 28 May. This initiated many disaster related plans for the AJG.[2]

During this time, news of a shortage of geography teaching resources in devastated schools reached members of the AJG Commission of Geographical Education (CGE). In response, authorized textbooks and atlases were sent to all elementary and junior high school students for free as they are vital learning resources for students. This relief effort was aided by the government, who supplied textbooks via local editorial boards and schools. Since the school atlas is used only from the fourth through seventh grade, this caused confusion at some local educational boards and schools regarding the allocation of free atlases. Other shortages included damaged geography teaching materials such as world globes, wall maps and photographic collections.

In response, the CGE put forward an "Emergency Proposal for Equipment of Geography Resources in Order to Reopen Geography/Social Studies School Lessons". This proposal, in two parts, was published by the AJG executive committee on 11 April 2011. The first component was a request to the government, local educational boards and schools that atlases should be re-supplied to all affected students, in the same manner as other types of textbooks. The second component proposed that key geography teaching materials be supplied to all. For the latter request, the AJG started a fundraising campaign called "Fundraising for Geography Education Restoration after the Great East Japan Earthquake Disaster: Geography Teaching Resources for Affected Schools" in cooperation with the Geographic Education Society of Japan (GESJ) and Japan Association of Geographical Space (JAGS). This fundraising was supported not only by members of these organizations, but also from overseas through the Commission on Geographical Education (CGE) of the International Geographical Union (IGU). The donations allowed the AJG to grant 5,000,000 JPY (approximately $50,000 USD) to 45 affected schools for geography education support.[3] Furthermore, the Japan Map Center (JMS) visited affected schools and supplied maps for instruction.

[2]The Disaster Response Headquarters was closed at the end of March 2012, but the outcomes are shown on the AJG website: http://www.ajg.or.jp/disaster/201103_Tohoku-eq.html.

[3]As stated in Sect. 20.3, outcomes from our program granted by TGS were used to select these schools.

20.5 Current State of Japanese and Geography Education

20.5.1 Government Education Policy After the Disaster: A National Dimension

According to Japanese education policy, disaster prevention education has a place in the school security education sphere. That policy states that disaster prevention education, belonging to Health and Physical Education subject as well as in a special school activity, should be practiced in close cooperation with other subjects. The disaster prevention education consists of three fields, life safety, traffic safety, and disaster prevention. Geography and Social Studies have contributed a great deal to this field of education, along with Health and Physical Education, Home Economics and Science. For example, the National Curriculum Standards prescribes the following units of a disaster prevention education.

Elementary School (Social Studies)
Third–fourth grade: disaster and accident prevention in the local area
Fifth grade: land preservation and natural disaster prevention

Junior High School (Geography Curriculum in Social Studies)
First–second grade: characteristics of the natural environment in Japan and disasters/disaster prevention

Senior High School (Geography)
First–third grade: natural environment and disaster prevention

This shows that geography education has carried out the disaster prevention education from the natural/human/social relationship viewpoint.

The Great East Japan Disaster demands reexamination of disaster prevention education policy in the government. The Ministry of Education, Culture, Sports, Science and Technology (MEXT) established the "Committee of Well-Informed Persons for Disaster Prevention Education After the Great East Japan Disaster Age" in July 2011. This committee produced a final report in July 2012 that recommended enrichment of disaster prevention education, especially fostering student's skills and attitudes about survival during hazards. The Cabinet of the government passed the "Plan of School Security Education Promotion" in April 2012. The promotion of disaster prevention education has been demanded at a national policy level. Geography and Social Studies are regarded as the main subjects to lead and achieve this aim.

20.5.2 Practices in Geography Education: A School Lesson Perspective

In school geography lessons, disaster prevention practices have certainly increased since the disaster. An examination of trends reported in academic association presentations and journals are detailed below.

20.5.2.1 Presentations at Academic Conferences

GESJ held their annual conference in August 2011, at Akita in northeast Japan. The deadline for the call for presentations was very soon after the disaster, so there was only one geography lesson plan presentation that addressed it. The presenter was a secondary school geography teacher in Sendai City, who suffered from the aftermath of the disaster alongside her students. In the conference, she first reported on damage to schools and students, and said geography education should confront this pressing situation with special care to affected students. She then proposed her geography lesson plan for third year of senior high school. This lesson effectively incorporated students' own disaster survival experiences into the Japanese high school geography curriculum.

In the subsequent GESJ annual conference in July 2012, there were three presentations that examined disaster prevention education. The themes were: (1) A theoretical framework for disaster prevention education in geography; (2) an analysis of lessons learned from the controversial issue of "radioactive contamination and region" to foster consensus building and social skills; and (3) a challenge to create a disaster-resilient city through fieldwork and map use. Disaster-related geography education presentations have also been increasing at AJG and Social Studies educators' conferences.

20.5.2.2 Publications in Journals

GESJ's academic journal *Shin Chiri* (The New Geography) published their 60th anniversary special issue on the "New Official Guideline for School Teaching and Strategy for Geography Education" in April 2012. This issue consists of various ideas for implementing the new National Curriculum Standards, with many references to disaster prevention education.

Kiyoshi Yoshikai, who was involved in this curriculum revision procedure at MEXT, stated that substantial disaster prevention education is an urgent issue for the government. He supports that geography be the core subject for promoting this, partly due togeography education's earlier achievements (GESJ 2012). In response, many researchers and educators discussed how geography contents and strategies contribute to disaster prevention education.

The monthly journal *Chiri* (Geography), published by the Kokon Shoin Corporation for geography readers such as school teachers, quickly released a special emergency issue in June 2011. The journal has since produced many articles and information concerning the disaster and affected areas. For example, a special issue titled "Japanese Society after the Disaster of 3.11" was published in January 2012. In this issue, a geography professor from an affected area described significant roles of geography education in disaster prevention education, through practices such as hazard map learning, awareness of local geography for survival, and workshops for raising disaster prevention consciousness (Murayama 2011). A group of high school geography teachers published the article "Geography Education Confronts Disasters" after the January 2012 issue, regarding their view that a disaster is a comprehensive social phenomenon that involves restoration.

20.5.2.3 Disaster Prevention, Education, and Geographical ESD Lessons

A review of events immediately following the Great East Japan Disaster presents many lessons regarding disasters and their processes that can foster student knowledge and skills for surviving similar events in the future. In 2005, the United Nations Decade of Education for Sustainable Development (UNDESD) had already stated that disaster prevention education was one of the ESD themes, making it clear that disaster prevention education is an important field in ESD.

Recently, geography lessons are increasingly concerned with ideal methods to restore devastated areas or ways to rebuild Japanese society resistant to disasters. These lessons are frequently discussed in Education for Sustainable Development (ESD).

ESD has recently been in demand in Japanese school education. It borrows from the theory and practice of geography, making geography a central and contributing subject. Productive examples of this synergy include the development of a book called *Jizoku kanou-na syakai to chiri kyouiku jissen* (Sustainable Society and Geography Education Practices) (Nakayama et al. 2011). It consists of ESD explanations, theoretical discussions about relationships between ESD and geography education, and proposals of Geography-ESD lesson plans from kindergarten through university. This publication is also considered a compilation of Japanese Geography-ESD research and practical achievements from the recent past to 2010; however, there are no lessons dealing with disasters.

One year after the disaster, the same editors published a guidebook for Geography-ESD lesson-making (Nakayama et al. 2012) that includes 15 recommended Geography-ESD lesson plans. Three of these are on disaster prevention, spanning kindergarten through university. For example, a lesson plan titled *Tunami-de shinsui shinakatta rekishi-kaido* (The Ancient Road Where Tsunami Never Reaches) considers tsunami damage in the Sendai plain from the 2011 disaster. If an ancient road map is overlaid on the inundated area map, it clearly shows that tsunamis never reach the ancient road. This suggests that ancient people applied their knowledge about the geographical environment and its associated hazards and risks to construct the road where it was safe from tsunamis. The plan proposes a Geography-ESD lesson in the hope that this discovery will inspire students to learn to apply disaster prevention practices.

More recently, in August 2012, a geography teacher group published a practical book on Geography and Social Studies related ESD lessons that reference the 2011 disaster. This publication is forward looking and attaches much importance to social participation. In particular, in a practice called *Nihon-no enerugii-mondai-wo kangaeru* (Thinking about the Energy Issue in Japan) (Izumi et al. 2012, pp. 41–49), students reflect on Japanese energy policy through mock national voting about nuclear power issues and then they evaluate the results. In the days before the 2011 disaster, few teachers would take up issues about nuclear disasters in Japan. However, there are real risks associated with nuclear power plants because they are located in tectonic zones like Japan, posing serious global issues (Crowley and Elliott 2012). Therefore, it is important to include the younger generation, who will one day be decision makers in the discussion of energy policy. The aforementioned lesson practice encourages geographically informed citizens to participate in creating a sustainable society, through discussion of key controversial issues.

20.6 Conclusions

The impact of the Great East Japan Disaster is not over. Approximately 316,000 people have been evacuated from their home and are still displaced across Japan (Reconstruction Agency in Cabinet publicity, at of 25 January 2013). In particular, it is unknown when or whether people who lived within a 20-km radius of the Fukushima-Daiichi Nuclear Power Plant will ever be able to return home, because radioactive contamination continues to pose a danger.

Under these circumstances, the geography education community is actively engaged in disaster education. For example, original and challenging lessons are being developed for examination at various spatial scales. At a prefectural scale, the Miyagi Prefectural Senior High School Geography Teacher Association revised their original supplementary textbook after the disaster, publishing a new edition in spring 2012. The new book includes substantial disaster prevention content that references recent local damages.

On a global scale, a U.S.–Japan collaboration developed geography teaching resources and brought participants on a fieldtrip to observe damage. The event was held in September 2012 and was sponsored by the Center for Global Geography Education of the Association of American Geographers, AJG and The University of Tokyo. Project members included Japanese and U.S. geography teachers and educators. The excursion to the affected areas provided inspiration and information for groups to collaboratively develop new resources, including the Great East Japan Disaster as a case study. These resources are available on the CGGE-AAG website.[4]

On the first anniversary of the Great East Japan Disaster, the AJG held a public symposium entitled "Think About Disasters and Education for the Young: the Present Situation of Devastated Schools and Future Geography".[5] The aim of this symposium considered the future of education, and there were presentations from educators, administrators (e.g., school superintendents and teachers in the affected area), as well as geographers. The symposium outcome positioned education as the most important aspect of constructing disaster prevention programs and for restoring a resilient society. The value of geography education was made clear, especially its key role in contributing to disaster education. Thus, geography education must have a vital role in the post Great East Japan Disaster era.[6]

[4] http://www.aag.org/cs/cgge/modules.

[5] *Chiri* published in May 2012 is the special issue for this symposium.

[6] After submission of this manuscript, Mitsuhashi (2013) was published. This paper summarizes the overall trend of Social Studies education in disaster prevention in Japan after the Great Japan Earthquake, and is therefore very helpful.

References

Crowley K, Elliott J (2012) Earthquake disasters and resilience in the global North: lessons from New Zealand and Japan. Geogr J 178(3):208–215

GESJ (2012) Tokushu: shin gakushu shido yoryo to "chiri" no sutorateji (New official guideline for school teaching and strategy for geography education: 60th anniversary special issue). Shin Chiri (New Geogr) 60(1)

Great East Japan Earthquake Taskforce Science Council of Japan (2011) Report to the foreign academies from science council of Japan on the Fukushima Daiichi nuclear power plant accident. Science Council of Japan, Tokyo

Izumi T, Umemura M, Fukushima Y, Ikeshita M (eds) (2012) Syakai-sankaku no jyugyou–zukuri; Jizoku-kanou na syakai ni mukete (Creating lessons as social participation: for sustainable society). Kokon shoin, Tokyo

Kurokawa, N. (2012). Higashi-nihon-daishinsai karano gakko no fukkou: Genjou to kadai (The recover from the Great East Japan Earthquake: the present state and agenda). Issue Brief, 736. National Diet Library

Mitsuhashi H (2013) Syakaika kyoiku ni okeru bousai kyoiku: Higashi Nihon Daishinsai go no gakkai shi ronbun tou wo chushin ni (Trends of disaster educational research in social studies education: after the Great Japan Earthquake). Syakaika kyouiku kenkyu (J Social Stud) 119:100–110

Murayama Y (2011) Gakko ni okeru bousai-kyoiku/ bousai- keikaku: Daishinnsai no keikenn wo fumaeta kenntou (School education and management for disaster prevention: based on experiences from the Great East Japan Earthquake and Tsunami). Chiri (Geogr) 57(1):40–45

Nakayama S, Wada F, Yuasa S (eds) (2011) Jizoku-kanou na syakai to chiri-kyouiku-jissen (Sustainable society and geography education practices). Kokon shoin, Tokyo

Nakayama S, Wada F, Yuasa S (eds) (2012) Jizoku-kanou na syakai wo mezasu chiri-ESD jyugyou gaido (Geography ESD lessons guidebook for sustainable society). Keibunsha, Onomichi

National Police Agency (2011) Tohoku-chiho taiheiyo-oki no higai-jokyo to keisatsu-sochi (The damage situation by the Tohoku-Pacific ocean earthquake and measure of national police agency). National Police Agency, Tokyo. http://www.npa.go.jp/archive/keibi/biki/higaijokyo.pdf. Accessed 20 Jan 2013

Obara K (2012) Donna Jishin dattanoka: Tohoku-chihou taiheiyou-oki jishin no tikyuu-kagaku-teki haikei, Gaiyou to eikyou (What kind of earthquake was it?: Earth scientific background, overview and influence of the Tohoku-Pacific Ocean Earthquake). In: Satake K, Hori M (eds) The sciences of the Great East Japan Earthquake. University of Tokyo Press, Tokyo, pp 155–189

Sawai Y, Shishikura M, Komatsubara J (2007) A study on paleotsunami using hand corer in Sendai plain (Sendai City, Natori City, Iwanuma City, Watari Town, Tamamoto Town), Miyagi, Japan. Ann Rep Active Fault Paleoearthquake Res 8:17–70

Tanaka H (2012) Higashi-nihon-daishinsai no keizaigaku-teki sokumen: Keizai kouzou henka to zaiseinan no Nihon wo haikei ni (The economic aspects of the Great East Japan Earthquake: against the economic structure change and financial difficulty). In: Satake K, Hori M (eds) The sciences of the Great East Japan Earthquake. University of Tokyo Press, Tokyo, pp 155–189

Tsunami Damage Mapping Team, Association of Japanese Geographers (2011) Maps of the Area hit by the Tsunami of 11 March 2011, Northeast Japan. http://danso.env.nagoya-u.ac.jp/20110311/map/index_e.html. Accessed 20 Jan 2013

Chapter 21
Basic Geography and the Future of Geography Education in Japan

Yoshiyasu Ida and Yumiko Takizawa

Abstract In Japan, many senior high schools require students to take a course in either Japanese History or Geography; most students select Japanese History. Consequently, many students fail to master the minimum level of basic geographic knowledge. To resolve this knowledge gap, the Science Council of Japan introduced in 2011 a plan to make Basic Geography and Basic History compulsory subjects. Basic Geography is designed as a compulsory two-credit class with the content structured from two perspectives, global-scale and local-scale. The course is developed around knowledge and skills that could help students become interested in current international issues, and the course is expected to create solutions from their independent ways of understanding and thinking about the world.

Keywords Basic Geography • Compulsory subject • Future • Teaching geography

21.1 Introduction: Background of Basic Geography

At present, only World History at the senior high school level, from the subject areas of Geography and History, is designated as compulsory. Many senior high schools require that students take either Japanese History or Geography, and most students' select Japanese History. However, a number of Japanese schools don't teach geography courses at all. Consequently, students fail to master basic geographic knowledge, made worse because few students learn Earth science. The growing trend is that increasingly more students graduate without learning about natural geography, the Earth, and the environment. The number of students who do not complete a geography course is growing, which contributes to young peoples' minimal understanding of and interest in maps and poor spatial perception, which are all skills needed to excel at World History and Japanese History.

Y. Ida (✉)
University of Tsukuba, 1-1-1 Tennodai, Tsukuba, Ibaraki 305-8572, Japan
e-mail: ida@human.tsukuba.ac.jp

Y. Takizawa
Teikyo University, Tokyo, Japan

Y. Ida et al. (eds.), *Geography Education in Japan*, International Perspectives in Geography 3, DOI 10.1007/978-4-431-54953-6_21

Currently, the number of students who study Geography is roughly half of all high school students. This wasn't always the case until Contemporary Society was designated as a compulsory social science subject in 1978. Prior to this period, almost all first-year senior high school students studied Geography, however, the enrollment numbers declined sharply after the policy change. Another education policy revision followed in 1989 in which Social Studies was divided into two categories: (1) Civics and (2) Geography and History. World History remained a compulsory subject, while Japanese History and Geography remained elective. Students could choose to study one of the two options, which further reduced the enrollment in Geography. Geography is again under threat as local governments, including that of the Tokyo Metropolis and the Kanagawa Prefecture, have recently proposed a plan to make Japanese History a compulsory subject in senior high schools. This is directly related to the low enrollment in geography. Despite the increasing importance of understanding international issues, global environmental issues, the mechanisms of natural disasters, and disaster education, geography is neglected in the Japanese school system. This is a problem that must be rectified promptly.

In 2006, 'the problem of students not studying World History' was taken up by the mass media, and it became a big issue. The Science Council of Japan set up the Subcommittee on Geography and History Education Senior High School in May 2007 to develop solutions. The committees for history, regional studies, geography and pedagogy were combined and governed by the membership and led by a chairperson (world history specialist), a deputy committee chairperson (geography specialist), a secretary (Japanese history specialist), as well as committee members specializing in pedagogy and archaeology/anthropology. Each field had an equal number of committee members represented.

To begin, the council recognized 'education with the right balance of temporal awareness and spatial awareness' as the guiding principal. World History has been a compulsory subject for over 20 years, and Geography has been relegated to elective status equally as long. The challenge at the moment is that a growing number of Japanese lack a minimum level of geographic knowledge. Many factors contribute to this, for example an increasing number of schools either do not teach geography or only teach it to students who aspire to study science. There is a disproportionate focus on passing university entrance examinations, and teachers spend little time teaching geography. To change this situation, the premise of education with the right balance of temporal and spatial awareness is especially important, and this has remained the focus throughout the subcommittee's activities (Yui 2011).

Initially, there were strong opinions about the combination of geography and history into a single subject (Sakurai 2011). Over meetings at two symposia and several exploratory investigations, the committee decided to create a consolidated reform plan, whereby two subjects—Basic History and Basic Geography—would be established as compulsory courses. The plan was proposed to the Science Council of Japan in 2011.

The proposal of Basic History would resolve the fragmentation of World History and Japanese History as it currently exists in the senior high school history curriculum. This change would also integrate Japanese History into World History, within which Global History is taught. It was also decided that the Basic History curriculum would reduce the traditional Eurocentric focus to include the origin and history of various world nations as well as their international activities. Instructional techniques expanded to include theme-oriented learning, investigative learning, and the collection and interpretation of historical sources, as an effort to cultivate historical thinking abilities. After numerous planning meetings to design the Basic History content, three draft plans have been submitted for future examination. The general ideas for each plan are laid out as Plan A: A chronological sequence model plus a theme-oriented learning model; Plan B: A focus on modern history; and Plan C: A theme formation model, as mentioned by Takahashi (2011).

The Examinations of Basic Geography were led by the AJG-CGE before the 2007 subcommittee was inaugurated. Similarly, examinations were also conducted by the School Education Subcommittee under the Joint Geography Education Subcommittee between the Area Studies Committee, the Science Council of Japan, and the Earth and Planetary Sciences Committee. There was also a plan to include more topography content in the Basic Geography curriculum, but after further discussions, it was decided that all ideas should be integrated into one plan, detailed in the next paragraph. Compared to the field of history, the geography community feels a strong sense of crisis on the state of geography education.

21.2 The Concept of Basic Geography

Geography as a subject in education has been criticized for being "the geography of place names and produce." This is because there had been a focus on understanding phenomena by rote learning, while the real essence of geography, namely, consideration of the relationships between nature and human beings were lacking in school education. However, in order to consider these interlocking ties, one must first have knowledge. Without knowledge, one will lack the resources for such consideration. What degree of geographic knowledge is required to consider the real essence of geography? Relentlessly pursuing such knowledge would lead to limitless investigation, and school education would end up being focused only on acquiring information. Applying the knowledge learned in school to resolve world issues can be done only after graduating and going out into the world itself. The foundation of geography is not simply to gain knowledge. How one applies knowledge, considers, and applies it is part of Basic Geography.

The Basic Geography plan has allotted time for students to study and reflect on basic knowledge learned. Furthermore, skills such as how to read maps and geographical ways of seeing and thinking have also been included as fundamental elements. Appropriate content has thus been designed as compulsory subjects for high school geography. Moreover, rather than merely considering, it was decided that students would also need to ascertain the process through which knowledge and skills are learned and used as the basis for studying and reflecting upon geography, to put their learning into practice. For example, following the Great East Japan

Earthquake, the testimonies of disaster victims provided indispensable information about disaster prevention against earthquakes and tsunamis. In Japan, which is very prone to earthquakes, such disaster prevention information can help not only to save one's own life but also the lives of many others. In light of this, information concerning disaster prevention is also included in Basic Geography, along with geography skills such as the use of maps and GIS. To make it easier to apply such knowledge and skills in real life situations, this teaching method also places great importance on field research.

As stated above, the concept of Basic Geography is not simply about facts. It also considers the relationships between nature and humans. It also includes the learning process, which begins with gaining geographic perspectives and knowledge, such as geographic ways of seeing and thinking, and ends with application in daily life. Furthermore, the integration of content such as disaster prevention, which is a major issue in Japan, and the use of modern tools such as GIS will help students survive and thrive in contemporary society. This skill set will also help them cultivate their capacity to build a peaceful and democratic society (Usui 2011). Thus, it may be appropriate to make Basic Geography a compulsory subject in the senior high school Geography and History curricula. This will support not only preparation for university entrance examinations, but also all courses to be taken after graduation.

21.3 The Structure of Basic Geography

The structure of Basic Geography has been described by Ida (2011, 2012), and these form the basis of discussion about Basic Geography. The plan is to develop Basic Geography as a compulsory two-credit class. Its content is structured from two perspectives: global-scale and local-scale. A methodical geographic content is adopted to present content from a global perspective, to impart the principal knowledge and skills necessary to foster student interest in international issues, and to develop solutions based on independent views and thinking process. The intention of this course is to cultivate students' ability to find solutions to current national and international issues by building on their topographical knowledge and views developed in geography classes up to junior high school. In terms of local perspectives, the content includes learning outcomes about the community, a topic taught in junior high school, and simultaneously introduces a wider perspective that tackles various regional geographical issues (e.g., environmental, disaster prevention, and industry) in a probing and problem-solving manner. These locally based lessons are intended to foster students' knowledge and skills to help them become involved in society in a meaningful way. Instruction through these two perspectives make up the foundation of Basic Geography, complemented by content instruction that considers the interactive relationships between global and local perspectives.

Table 21.1 shows topics, subtopics, and the learning content of major topic I of the Basic Geography course. Tentatively, 70–80 % of class time will be allocated to major topic I, and 20–30 % to major topic II.

Table 21.1 Major topic I: Features and issues in the modern world

Topics	Subtopics	Keywords
1. Modern World from a Global Perspective	(1) Structure of the Earth	Earth, general circulation of atmosphere/oceans and weather/vegetation zones, plate tectonics and global-scale disasters, global environmental changes
	(2) Connected world	Global population movement, global flow of goods, information and funds, energy revolution and trade, international organisations, national territory and territorial waters, information and telecommunications and the digital Earth
2. Diverse Natural Environments and Cultures of Different Regions	(1) Diversity in the natural environment found in various regions	Mobile zones/stable zones, from tropics to polar regions, arid zones, monsoon regions, mountainous regions/plains/islands
	(2) Diversity in lifestyle/culture found in various regions	People, religions, languages, clothing / diet / dwelling
	(3) Diversity in the relationships between nature and lifestyle/culture found in various regions	Disasters/environmental pollution/destruction and human activities
3. Global Economy/Society and Various Issues	(1) Global economic system	Blocking and multinational industries, environment and economy, central regions and peripheral regions
	(1) Population/agriculture/food	The population rise and the declining birth rate and aging population, agriculture and agricultural regions, food systems and agribusiness
	(2) Industrialization and regional disparity	Resources and location, industrialization and industrial regions, de-industrialization and industrial clusters, new international specialization
	(3) Urban issues	Urban systems and mega cities (global cities), multiculturalism, poverty, cities and disasters, cities and the environment, transportation infrastructure

Major topic I contains three topics: (1) a global perspective; (2) diverse natural environments and cultures of different regions; and (3) the global economy and society and various issues. In studying the modern world from a global perspective, students are presented with a general view of the world by studying the different ways that people are connected through globalization. By way of learning about the diverse natural environments and cultures of different regions, students will be able to look at the world from the three perspectives: (1) the natural environment, (2) lifestyles and cultures, and (3) the relationships between nature and culture. Case studies that examine regions from a range of perspectives are used to demonstrate the continuum of issues in the global economy and society.

Table 21.2 Major topic II: Features of communities and local area development

Topics	Learning Content
1. Features of communities	The natural environment and land use in regions, regional population, industry, and history
2. Various local geographic issues and local area development	A. Regional development that seeks preservation and use of natural resources in local area
	B. Regional development that seeks to prepare against natural disasters and alleviate damage in local area
	C. Regional development that seeks preservation and use of historical resources in local area
	D. Regional development that seeks the improvement of the environment and water circulation in local area
	E. Regional development that seeks the enhancement of production, employment, and sales service in local area
	F. Regional development that seeks the enhancement of health and welfare measures in local area
	G. Regional development that seeks efficient use of energy in local area
	H. Regional development that seeks to make transport in the region more comfortable in local area
	I. Regional development that seeks environmental improvement to protect against accidents and crime in local area

Major topic II consists of two topics: (1) the features of communities and (2) various local geographic issues and local area development. Table 21.2 outlines major topic II which features community and local area content that enables students to investigate their surrounding area by probing and using problem-solving strategies (e.g., inquiry, data collection). In the process, students discover diverse issues related to geographic features, the environment, disasters, and industries within the local region where they conduct field research. Moreover, students begin to link these experiences with the learning outcomes of major topic I. Together, these major topics lead student development of knowledge and skills that are valuable if they participate in community building.

Major topic II is designed to help students understand the features and issues of a community from a local scale and to be mindful about the connections with the learning outcomes of major topic I. In the first topic, 'features of communities', materials such as traditional maps, as well as electronic maps and satellite images acquired from Web GIS, will be used as learning tools to discover the basic characteristics of the local community where students live. The second topic, 'various local regional issues and local area development' builds upon such regional characteristics, and, while conducting field research, students are expected to learn about local geographical issues related to the natural environment, disasters, and industry. The next step to extend this learning so students can use the inquiry approach to more deeply understand the problems and to suggest potential solutions.

If all senior high school students study and build on the learning outcomes of Basic Geography—and take elective courses (e.g., Geography B)—which comprise detailed world topography or more in-depth physical geographic content, they will gain a deeper understanding of Japan and the world. Moreover, students will be able to understand the value of learning geography.

21.4 Characteristics and Issues of Basic Geography

The characteristics of Basic Geography are foundation to the field. These are summarized in the four points below:

(1) Considering current and important geographic phenomena,
(2) Giving close consideration to place, space, and time,
(3) Presenting the natural environment and social/cultural issues in a comprehensive and integrated manner by undertaking specific case studies,
(4) Conducting examinations by employing maps and GIS.

Geography education plays a role in ESD, which fosters individuals who are conscious of their relationship with others, their relationship with society, their relationship with the natural environment, and who are able to highly value relations and connections. Basic Geography offers content that is relevant to these ESD concepts.

There are two issues worth mentioning about the implementation of Basic Geography. The first point is that since Basic Geography is a two-credit course, the content that forms the "basics" is carefully selected. There is no consensus among geographers on the elements of geography; however, it is aligned to the geography curriculum taught throughout school grades. Geography education researchers in universities, geography teachers in senior high schools, and junior high school Social Science teachers who specialize in geography contributed to its formulation and reviewed the ensuing drafts. Geography researchers were also involved as readers of the content and provided suggested revisions.

The second point is whether Basic Geography will be taught by all geography and history educators. It would be optimal for Basic Geography to be taught by teachers specializing in geography; however, owing to a shortage of trained geography teachers, it is conceivable that teachers specializing in history will teach it. The obstacle for non-geography teachers is that the knowledge, skills, and ways of thinking in Basic Geography are not simple. This issue has to do with the way that the university teaching certificate and in-service training systems are organized. Social teachers have a license that allows them to teach geography and history. Unfortunately, many teachers did not major in geography at university and their last geography class was in junior high school. In contrast, many more teachers have studied history in senior high school. Even if teachers have no experience in geography, they could be allocated the class. This causes geography education supporters to worry about the state of geography in schools. The lesson is that in-service training for teachers is important to implementing content rich courses.

21.5 Toward the Implementation of Basic Geography

The Science Council of Japan has established the aims and general framework of Basic Geography; however, other course elements, such as specific content items, are still in draft form. This chapter summarizes the specific learning content based on this draft plan and the challenges of implementing these plans are outlined below.

The first challenge is to verify the impact of the Basic Geography curriculum. If a large number of schools implement this course in their capacity as designated development schools, there is a high possibility that Basic Geography will be taught as a compulsory subject in senior high schools. This is conditional upon the success of the first implementation phase before it is included in the next National Curriculum Standards revision.

The next issue concerns school credits. The total credits for Basic Geography and Basic History courses are four. Under present regulations, the minimum number of credits to be taken is four, depending upon the school, students will only be able to take compulsory subjects. Many senior high schools will probably teach elective geography as well as compulsory subjects, but in schools where only four credits of Geography and History are taught, students will have no freedom to select subjects. The challenge is whether credit rules will change in the favor of Basic Geography.

The third concern is how to integrate technology into the classrooms. Basic Geography recognizes the importance of maps and GIS, but how the geography education community will impart GIS skills in schools is a real challenge.

Despite these challenges, Basic Geography promises a comprehensive course that fosters students' basic knowledge of geographical phenomena, ways of viewing and thinking about the world, and geographical skills that are critical to making rational judgments and decision making for a peaceful and promising society.

References

Ida Y (2011) Koko chiri rekishi ka hisshu kamoku toshiteno 'chiri kiso' an (The content of basic geography presented by SCJ). Gakujutsu no doko (Trends Sci) 186:28–34

Ida Y (2012) Kotogakko chiri rekishi ka hisshuka ni mukete (Compulsory subject of geography in high school). Shin Chiri (New Geogr) 60(1):77–79

Sakurai Y (2011) GIS to chireki togo (GIS and the integration with geography and history). Gakujutsu no doko (Trends Sci) 186:56–59

Takahashi M (2011) Shinkamoku 'rekishikiso' no tokucho to gutaika ni mukete (The characteristics and realization of basic history). Gakujutsu no doko (Trends Sci) 186:22–27

Usui T (2011) Dejitaruka to chirikyoiku (Digitization and geography education). Gakujutsu no doko (Trends Sci) 186:52–55

Yui D (2011) Koko chiri rekishika kyoiku no genjo to kaikakuan no zentaizo (The actual situation of geography and history, and outline of revaluation for the subject in high school). Gakujutsu no doko (Trends Sci) 186:8–13

Conclusion

The Japanese geography education systems has emphasized descriptive regional geography at the school level. However, efforts are made towards fostering students' skills, attitudes, and interest in GIS and map use, active field research, social participation through geography integration in education for sustainable development (ESD), and disaster preparedness. These changes parallel improvements made to the teacher training system for developing well-qualified geography teachers. These actions are confronted by both institutional and societal challenges such as the general public's impression that geography is not important.

Since geography's change from a compulsory to elective status, negative impacts are seen in senior high schools. Thus, to demonstrate geography's value to society, the current geography curriculum in senior high schools is being redeveloped and experimental courses piloted in the classrooms. This provides a model for making geography compulsory in senior high school. Compulsory geography at this level would contribute greatly to the development of geography education in Japan by demonstrating its value to society and that geography is an important part of a citizen's education.

International research findings are currently being considered for the Japanese geography education system. The work ahead is to firmly establish that geography education and curricular designs meet world standards as well as to ensure that teacher training programs are apace with the research and developments in high performing countries worldwide. With collaborative and independent efforts, changes are emerging. For example, through participation in the International Geography Olympiad, students and many geography teachers' views are changing and leading to novel experiences and progress. While they are following Japan's traditional geography education system, exploration along a new path is underway. This change in geography education is unique to Japan but also incorporates world geography standards. We look forward to more developments in Japanese geography education that build on the aforementioned efforts, both planned and already in progress.

Yoshiyasu Ida

Y. Ida et al. (eds.), *Geography Education in Japan*, International Perspectives in Geography 3, DOI 10.1007/978-4-431-54953-6

MIX
Papier aus verantwortungsvollen Quellen
Paper from responsible sources
FSC® C105338

If you have any concerns about our products,
you can contact us on
ProductSafety@springernature.com

In case Publisher is established outside the EU,
the EU authorized representative is:
Springer Nature Customer Service Center GmbH
Europaplatz 3, 69115 Heidelberg, Germany

Printed by Libri Plureos GmbH
in Hamburg, Germany